T0074189

STEAM-H: Science, Technology, Engineering, Agriculture, Mathematics & Health

STEAM-H: Science, Technology, Engineering, Agriculture, Mathematics & Health

Series Editor
Bourama Toni
Department of Mathematics
Howard University
Washington, DC, USA

This interdisciplinary series highlights the wealth of recent advances in the pure and applied sciences made by researchers collaborating between fields where mathematics is a core focus. As we continue to make fundamental advances in various scientific disciplines, the most powerful applications will increasingly be revealed by an interdisciplinary approach. This series serves as a catalyst for these researchers to develop novel applications of, and approaches to, the mathematical sciences. As such, we expect this series to become a national and international reference in STEAM-H education and research.

Interdisciplinary by design, the series focuses largely on scientists and mathematicians developing novel methodologies and research techniques that have benefits beyond a single community. This approach seeks to connect researchers from across the globe, united in the common language of the mathematical sciences. Thus, volumes in this series are suitable for both students and researchers in a variety of interdisciplinary fields, such as: mathematics as it applies to engineering; physical chemistry and material sciences; environmental, health, behavioral and life sciences; nanotechnology and robotics; computational and data sciences; signal/image processing and machine learning; finance, economics, operations research, and game theory.

The series originated from the weekly yearlong STEAM-H Lecture series at Virginia State University featuring world-class experts in a dynamic forum. Contributions reflected the most recent advances in scientific knowledge and were delivered in a standardized, self-contained and pedagogically-oriented manner to a multidisciplinary audience of faculty and students with the objective of fostering student interest and participation in the STEAM-H disciplines as well as fostering interdisciplinary collaborative research. The series strongly advocates multidisciplinary collaboration with the goal to generate new interdisciplinary holistic approaches, instruments and models, including new knowledge, and to transcend scientific boundaries.

More information about this series at http://www.springer.com/series/15560

Anthony A. Ruffa • Bourama Toni
Editors

Advanced Research in Naval Engineering

Editors
Anthony A. Ruffa
Naval Undersea Warfare Center
Newport, RI, USA

Bourama Toni
Department of Mathematics
Howard University
Washington, DC, USA

ISSN 2520-193X ISSN 2520-1948 (electronic)
STEAM-H: Science, Technology, Engineering, Agriculture, Mathematics & Health
ISBN 978-3-319-95116-4 ISBN 978-3-319-95117-1 (eBook)
https://doi.org/10.1007/978-3-319-95117-1

Library of Congress Control Number: 2018956303

Mathematics Subject Classification: 30E25, 37M10, 40C05, 42B10, 65D07, 62C10, 62P30, 68T40, 76D09, 91A40

This Springer imprint is published by the registered company Springer Nature Switzerland AG
The registered company address is: Gewerbestrasse 11, 6330 Cham, Switzerland

Preface

The multidisciplinary STEAM-H series (Science, Technology, Engineering, Agriculture, Mathematics, and Health) brings together leading researchers to present their work in the perspective to advance their specific fields and in a way to generate a genuine interdisciplinary interaction transcending disciplinary boundaries. All chapters therein were carefully edited and peer-reviewed; they are reasonably self-contained and pedagogically exposed for a multidisciplinary readership.

Contributions are by invitation only and reflect the most recent advances delivered in a high standard, self-contained way. The goals of the series are as follows:

1. To enhance multidisciplinary understanding between the disciplines by showing how some new advances in a particular discipline can be of interest to the other discipline, or how different disciplines contribute to a better understanding of a relevant issue at the interface of mathematics and the sciences.
2. To promote the spirit of inquiry so characteristic of mathematics for the advances of the natural, physical, and behavioral sciences by featuring leading experts and outstanding presenters.
3. To encourage diversity in the readers' background and expertise, while at the same time structurally fostering genuine interdisciplinary interactions and networking.

Current disciplinary boundaries do not encourage effective interactions between scientists; researchers from different fields usually occupy different buildings, publish in journals specific to their field, and attend different scientific meetings. Existing scientific meetings usually fall into either small gatherings specializing on specific questions, targeting specific and small group of scientists already aware of each other's work and potentially collaborating, or large meetings covering a wide field and targeting a diverse group of scientists but usually not allowing specific interactions to develop due to their large size and a crowded program. Traditional departmental seminars are becoming so technical as to be largely inaccessible to anyone who did not coauthor the research being presented. Here contributors focus on how to make their work intelligible and accessible to a diverse audience, which in the process enforces mastery of their own field of expertise.

This volume, as the previous ones, strongly advocates multidisciplinarity with the goal to generate new interdisciplinary approaches, instruments, and models including new knowledge, transcending scientific boundaries to adopt a more holistic approach. For instance, it should be acknowledged, following Nobel laureate and president of the UK's Royal Society of Chemistry, Professor Sir Harry Kroto, "that the traditional chemistry, physics, biology departmentalised university infrastructures—which are now clearly out-of-date and a serious hindrance to progress—must be replaced by new ones which actively foster the synergy inherent in multidisciplinarity." The National Institutes of Health and the Howard Hughes Medical Institute have strongly recommended that undergraduate biology education should incorporate mathematics, physics, chemistry, computer science, and engineering until "interdisciplinary thinking and work become second nature." Young physicists and chemists are encouraged to think about the opportunities waiting for them at the interface with the life sciences. Mathematics is playing an ever more important role in the physical and life sciences, engineering, and technology, blurring the boundaries between scientific disciplines.

The series is to be a reference of choice for established interdisciplinary scientists and mathematicians and a source of inspiration for a broad spectrum of researchers and research students, graduates, and postdoctoral fellows; the shared emphasis of these carefully selected and refereed contributed chapters is on important methods, research directions, and applications of analysis including within and beyond mathematics. As such, the volume promotes mathematical sciences, physical and life sciences, engineering, and technology education, as well as interdisciplinary, industrial, and academic genuine cooperation.

Toward such goals, the following chapters are featured in the current volume.

The chapter "Impedance of Pistons on a Two-Layer Medium with Inviscid Homogeneous Flow" by Scott Hassan uses an integral transform technique to develop a general solution for the impedance of two-dimensional pistons acting on a two-layer medium.

In the chapter "Acoustics of a Mixed Porosity Felt Airfoil," Aren Hellum presents experimental results to understand the relationship between wing porosity and noise reduction as a function of flow Reynolds number.

In the chapter "Generalizing the Butterfly Structure of the FFT," John Polcari develops the underlying structure of the fast Fourier transform (FFT) well described by a pattern of "butterfly" operations in a more general context of a decomposition applicable to any arbitrary complex unitary matrix.

The chapter "Development of an Aft Boundary Condition for a Horizontally Towed Flexible Cylinder" by Anthony A. Ruffa uses the method of characteristics to develop an aft boundary condition for the linear transverse dynamics of a towed neutrally buoyant flexible cylinder, extending a previous approach to more general numerical methods.

In the chapter "Tracking with Deterministic Batch Trackers," Steven Schoenecker develops two deterministic non-Bayesian batch trackers—the

Maximum Likelihood Probabilistic Data Association Tracker (ML-PDA) and the Maximum Likelihood Probabilistic Multi-Hypothesis Tracker (ML-PMHT), assuming a target with some unknown deterministic motion corrupted by measurement noise.

In chapter "Moving Horizons Estimation for Wheelchair Trajectory Repeatability in the Home" by Steven B. Skaar shows an advantage of Moving Horizons Estimation in contrast with the previously tested Extended Kalman Filter in the context of achieving a useful form of teach/repeat for wheelchairs of severely disabled veterans within their homes.

In the chapter "Exact Solutions to the Spline Equations," Anthony A. Ruffa and Bourama Toni develop the exact solutions to the cubic spline equations for the case of equal knot spacing, exhibiting an oscillatory response in the region of a discontinuity as a consequence of the row structure of the resulting tridiagonal Toeplitz system.

The chapter "Distributed Membership Games for Planning Sensor Networks" by Thomas Wettergen and Michael Traweek investigates the management of sensor networks organized into distinct groups that share data to improve performance, in particular, for systems where changes in individual sensors' group membership may improve performance.

In chapter "Statistical Models of Inertial Sensors and Integral Error Bounds," [Richard] Vaccaro and Ahmed Zaki consider the random components that are useful for modeling modern tactical grade MEMS sensors with the purpose of deriving formulas for bounding errors in the first and second integrals of the sensor output from the additive noise and random drift components that corrupt a sensor signal.

The chapter "Developing Efficient Random Flight Searches in Bounded Domains" by Thomas Wettergren develops a new method for computing the parameters defining an efficient random flight for searchers that are constrained to search in a bounded domain, using concepts from observations of animal foraging behavior to define a random search plan that provides an optimally efficient search in terms of coverage relative to the constraints of random motion in the bounded domain.

The book as a whole certainly enhances the overall objective of the series, that is, to foster the readership interest and enthusiasm in the STEAM-H disciplines (Science, Technology, Engineering, Agriculture, Mathematics, and Health), stimulate graduate and undergraduate research, and generate collaboration among researchers on a genuine interdisciplinary basis.

The STEAM-H series is hosted at Howard University, Washington DC, USA, an area that is socially, economically, intellectually very dynamic and home to some of the most important research centers in the USA. This series, by now well established and published by Springer, a world-renowned publisher, is expected to become a national and international reference in interdisciplinary education and research.

Newport, RI, USA — Anthony A. Ruffa
Washington, DC, USA — Bourama Toni

Acknowledgments

We would like to express our sincere appreciation to all the contributors and to all the anonymous referees for their professionalism. They all made this volume a reality for the greater benefit of the community of science, technology, engineering, agriculture, mathematics, and health.

Contents

Contributors

Scott E. Hassan Naval Undersea Warfare Center, Newport, RI, USA

Aren M. Hellum Naval Undersea Warfare Center, Newport, RI, USA

John Polcari Informative Interpretations LLC, Burke, VA, USA

Oak Ridge National Laboratory, Oak Ridge, TN, USA

Anthony A. Ruffa Naval Undersea Warfare Center, Newport, RI, USA

Steven Schoenecker Naval Undersea Warfare Center, Newport, RI, USA

Steven B. Skaar Department of Aerospace and Mechanical Engineering, University of Notre Dame, Notre Dame, IN, USA

Bourama Toni Department of Mathematics, Howard University, Washington, DC, USA

C. Michael Traweek Office of Naval Research, Arlington, VA, USA

Richard J. Vaccaro Department of Electrical, Computer, and Biomedical Engineering, University of Rhode Island, Kingston, RI, USA

Thomas A. Wettergren Naval Undersea Warfare Center, Newport, RI, USA

Ahmed S. Zaki Naval Undersea Warfare Center, Division Newport, Newport, RI, USA

Impedance of Pistons on a Two-Layer Medium with Inviscid Homogeneous Flow

Scott E. Hassan

1 Introduction

Predicting the impedance associated with a piston in a rigid baffle, acting on a two-layer medium with flow, is a fundamental problem in the area of acoustic transduction associated with high speed flows.

Much of the previous work has addressed various aspects of pistons in rigid infinite baffles acting directly on a semi-infinite quiescent fluid. The radiation impedance associated with fluid loaded circular pistons has been developed extensively using approaches involving surface integrals of Green's functions [1], integral transforms [2], or time domain techniques [3]. General calculation methodologies with specific application to a rectangular aperture in a rigid baffle were addressed by Pierce et al. [4].

Previous work related to the response of rigid pistons on viscoelastic or a layered viscoelastic medium are limited to computing the impedances for a layered medium without flow. An integral equation approach was developed by Luco [5, 6] and solved numerically for a circular piston on a layered viscoelastic medium. Hassan [7] developed an integral transform method to solve for the self and mutual impedances of pistons on a two-layer medium without flow. It was found that the low frequency piston impedance is significantly influenced as a result of shear properties of the viscoelastic medium.

The problem of a piston, in a rigid baffle, acting directly on a moving homogeneous fluid medium has been addressed by Ffowcs Williams and Lovely [8], Leppington and Levine [9], and Levine [10] for a vibrating strip.

S. E. Hassan (✉)
Naval Undersea Warfare Center, Newport, RI, USA
e-mail: scott.hassan@navy.mil

A. A. Ruffa, B. Toni (eds.), *Advanced Research in Naval Engineering*, STEAM-H: Science, Technology, Engineering, Agriculture, Mathematics & Health, https://doi.org/10.1007/978-3-319-95117-1_1

Numerical and experimental methods to evaluate the acoustic impedance of rectangular panels in a rigid baffle with uniform subsonic mean flow were considered by Chang and Leehey [11]. They noted significant discrepancies between numerical and experimental results for the modal radiation reactance. Transient and harmonic fluid loading on vibrating plates in a uniform flow field was addressed by Li and Stepanishen [12]. They noted an unexpected behavior at low frequencies where the reactance was found to be very large. This was attributed to the nonzero limiting value of the impulse responses for nonzero Mach numbers.

In the present paper, integral transform methods are used to develop a general solution for the impedance of pistons acting on a two-layer medium consisting of a semi-infinite acoustic fluid with homogeneous flow, on a viscoelastic solid in a rigid infinite baffle. The theory is presented in Sect. 2 where the stresses in the solid are determined using the theory of linear elasticity and the pressures in the fluid are assumed to be governed by the convected wave equation. The special case of a piston of length L is considered in detail. Results for the piston impedance for various viscoelastic layer thicknesses and mean flow velocities are presented in Sect. 3 along with limiting classical solutions.

2 Theory

In this section, integral transform techniques are utilized to develop an expression for the impedance of a piston. The impedance is expressed as an integral containing the product of transformed aperture functions and the spectral impedance of a general multi-layer medium. The governing differential equations of motion are then developed for the two layers consisting of a viscoelastic solid and inviscid fluid with homogeneous subsonic flow. The equations of linear elasticity are used to describe the response of the isotropic viscoelastic medium, and the convected acoustic wave equation is used to describe the response of the fluid. These equations are coupled by applying boundary conditions at the fluid/solid interface and rigid baffle resulting in a system of linear equations for the unknown coefficients to the displacement potential functions. These coefficients are used to evaluate the stresses acting on the rigid baffle and the associated spectral impedance of the two-layer medium.

2.1 Formulation of Impedance Expressions

Consider the inhomogeneous two-layer medium depicted in Fig. 1. A semi-infinite acoustic fluid volume, V_f, with homogeneous subsonic mean flow is above an isotropic viscoelastic solid volume, V_s, of thickness h. The lower surface of the viscoelastic solid is on a rigid infinite planar baffle at $z = 0$. The piston described by σ is vibrating with a harmonic ($e^{j\omega t}$) uniform velocity. As a result of the piston motion normal to the baffle, a net normal force is exerted on the surface σ.

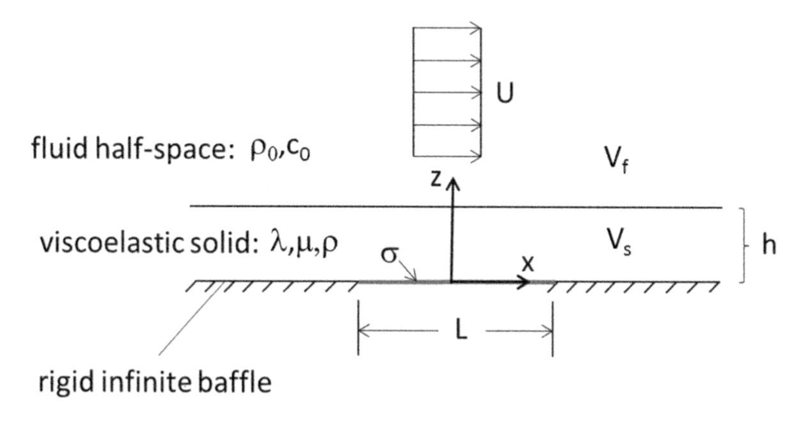

Fig. 1 Piston acting on a two-layer medium in a rigid infinite planar baffle

The radiation impedance, z(ω), is defined as the total normal force acting on the piston normalized by the complex normal velocity of the piston, V(ω). It follows that:

$$z(\omega) = \frac{F(\omega)}{V(\omega)} \tag{1}$$

where F(ω) is the complex normal component of force obtained by integration of the normal stress component on $z = 0$ over the surface σ. The general expression for the impedance can also be written as:

$$z(\omega) = \left[R(\omega) + jX(\omega)\right] \tag{2}$$

where R and X are the resistive and reactive components of the complex impedance. In the subsequent development, explicit dependence on $e^{j\omega t}$ is suppressed for convenience.

The solution to the problem of interest is pursued via the use of the following Fourier transform pair:

$$\hat{b}(k_x) = \int_{-\infty}^{\infty} b(x)\, e^{-jk_x x} dx, \tag{3a}$$

$$b(x) = \frac{1}{2\pi} \int_{-\infty}^{\infty} \hat{b}(k_x)\, e^{jk_x x} dk_x, \tag{3b}$$

where b(x) is any field variable of interest and the hat indicates a transformed quantity.

The complex impedance in Eq. (1) can be expressed as an integral of the spectral impedance weighted by a transformed aperture function [7]. It follows that:

$$z = \frac{1}{2\pi} \int_{-\infty}^{\infty} \hat{Z}(k_x) \left| \hat{S}(k_x) \right|^2 dk_x \tag{4}$$

where $\hat{Z}(k_x)$ is the spectral impedance relating the transformed normal stress to the transformed normal velocity. This quantity is expressed as:

$$\hat{Z}(k_x) = -\left. \frac{\hat{\tau}_{zz}(k_x, z)}{\hat{v}(k_x, z)} \right|_{z=0} \tag{5}$$

where $\hat{\tau}_{zz}(k_x, z)$ and $\hat{v}(k_x, z)$ are the transformed normal stress and velocity, respectively.

The S(x) is the real-valued aperture function for surface σ, and $\hat{S}(k_x)$ is the transformed aperture function. It is noted that expressions similar to Eq. (4) have also been used for the classical fluid-loaded rectangular piston on a planar infinite rigid baffle [13].

2.2 *Aperture Functions*

The transformed aperture functions in Eq. (4) act as wavenumber filters to the spectral impedance associated with the two-layer medium. The filter characteristics are a function of the aperture geometric parameters and the weighting function. For the present study, the transformed aperture function is associated with the region over which the normal stress on the piston is integrated. For the case of a rigid piston on a planar infinite baffle acting on a linear acoustic fluid, a unit weighting over the aperture is implied. This results in a discontinuous aperture function and an associated discontinuous velocity boundary condition. The velocity field in the viscoelastic solid is continuous and therefore requires a continuous aperture function. Continuous aperture functions have been considered by Tjotta and Tjotta [14] for the study of acoustic fields due a piston with non-uniform velocity distribution. The piecewise linear aperture function used for the present study provides the simplest transition in the small peripheral region, between a uniform normal piston velocity and a rigid baffle. Additionally, for small δ, results from the present study can be compared to classical baffled piston impedance functions.

It follows that the aperture function S(x) can be defined for the special case of a piston of length L with a continuous weighting. This aperture function is expressed as:

$$S(x) = \left\{ \begin{array}{cc} 1; & |x| \le \frac{L}{2} \\ \left(1 + \frac{L}{2\delta}\right) - \frac{x}{\delta}; & \frac{L}{2} \le x \le \left(\frac{L}{2} + \delta\right) \\ \left(1 + \frac{L}{2\delta}\right) + \frac{x}{\delta}; & -\left(\frac{L}{2} + \delta\right) \le x \le -\frac{L}{2} \\ 0; & \text{otherwise} \end{array} \right\} \tag{6}$$

The parameter δ in Eq. (6) controls the width of the peripheral region of the piston where the weighting function varies linearly from unity to zero. It is typical that $\delta/L \ll 1$ and is selected at $\delta/L = 0.02$ for the present study. The transformed aperture function is found by substituting Eq. (6) into Eq. (3a):

$$\hat{S}(k_x) = \frac{2}{\delta k_x^2}\left\{\cos\left(\frac{k_x L}{2}\right) - \cos\left(\frac{k_x L}{2} + k_x\delta\right)\right\}. \tag{7}$$

The transformed aperture function for a rigid piston of length L is obtained from Eq. (7) as a limiting case where $\delta \to 0$. It follows that:

$$\hat{S}(k_x) = L\frac{\sin(k_x L/2)}{k_x L/2}. \tag{8}$$

2.3 *Impedance Expressions for the Pistons*

The impedance of a piston acting on the two-layer medium is obtained by substituting Eq. (7) into Eq. (4) resulting in the following:

$$z = \frac{2}{\pi\delta^2}\int_{-\infty}^{\infty} \hat{Z}(k_x)\frac{\{\cos(k_x L/2) - \cos(k_x L/2 + k_x\delta)\}^2}{k_x^4}dk_x \tag{9}$$

The classical piston impedance expression is obtained from Eq. (9) as a limiting case where $\delta \to 0$. It follows that:

$$z = \frac{L^2}{2\pi}\int_{-\infty}^{\infty} \hat{Z}(k_x)\left\{\frac{\sin(k_x L/2)}{k_x L/2}\right\}^2 dk_x \tag{10}$$

Analytical solutions to the classical piston impedance functions associated with an acoustic fluid, obtained using integral transform methods, can be found in standard texts [2, 13].

2.4 Equations of Motion for the Viscoelastic Solid

The irrotational (ϕ) and rotational (ψ) displacement potentials for the viscoelastic solid satisfy the following reduced wave equations:

$$\nabla^2 \phi(\mathbf{x}) + k_d^2 \phi(\mathbf{x}) = 0, \tag{11}$$

$$\nabla^2 \psi(\mathbf{x}) + k_s^2 \psi(\mathbf{x}) = 0, \tag{12}$$

where ∇^2 is the Laplacian operator in the cartesian coordinate system with $\mathbf{x} = (x, z)$ and the complex dilatational and shear wavenumbers are indicated by the subscripts. The dilatation and shear wave speeds are defined in a standard manner using complex Lamé coefficients, μ and λ to account for dissipation in the viscoelastic medium. Substituting Eqs. (11) and (12) into Eq. (3a) results in the following set of transformed equations for the displacement potentials:

$$\frac{\partial^2 \hat{\phi}(k_x, z)}{\partial z^2} + \gamma_d^2 \hat{\phi}(k_x, z) = 0 \tag{13}$$

$$\frac{\partial^2 \hat{\psi}(k_x, z)}{\partial z^2} + \gamma_s^2 \hat{\psi}(k_x, z) = 0 \tag{14}$$

The solution to the above equations can be expressed as:

$$\hat{\phi}(k_x, z) = \hat{A}_1(k_x)\sin(\gamma_d z) + \hat{B}_1(k_x)\cos(\gamma_d z) \tag{15}$$

$$\hat{\psi}(k_x, z) = \hat{A}_2(k_x)\sin(\gamma_s z) + \hat{B}_2(k_x)\cos(\gamma_s z) \tag{16}$$

The transformed displacements in the viscoelastic medium are found using a standard approach [15] that involves expressing the displacements as a linear combination of a scalar and vector potential. It follows that:

$$\begin{aligned} \hat{u}_x(k_x, z) = jk_x &\left[\hat{A}_1(k_x)\sin(\gamma_d z) + \hat{B}_1(k_x)\cos(\gamma_d z)\right] - \\ \gamma_s &\left[\hat{A}_2(k_x)\cos(\gamma_s z) - \hat{B}_2(k_x)\sin(\gamma_s z)\right] \end{aligned} \tag{17}$$

$$\begin{aligned}\hat{u}_z(k_x, z) = \gamma_d \left[\hat{A}_1(k_x)\cos(\gamma_d z) - \hat{B}_1(k_x)\sin(\gamma_d z)\right] + \\ jk_x\left[\hat{A}_2(k_x)\sin(\gamma_s z) + \hat{B}_2(k_x)\cos(\gamma_s z)\right]\end{aligned} \tag{18}$$

The transformed stresses are now obtained using the standard linear constitutive and strain–displacement relationships [15] for the plane strain problem of interest. The final result is

$$\begin{aligned}\hat{\tau}_{zz}(k_x, z) = -\left(\lambda k_d^2 + 2\mu\gamma_d^2\right)\left[\hat{A}_1(k_x)\sin(\gamma_d z) + \hat{B}_1(k_x)\cos(\gamma_d z)\right] + \\ 2j\mu k_x\gamma_s\left[\hat{A}_2(k_x)\cos(\gamma_s z) - \hat{B}_2(k_x)\sin(\gamma_s z)\right]\end{aligned} \tag{19}$$

$$\begin{aligned}\hat{\tau}_{xz}(k_x, z) = 2j\mu k_x\gamma_d\left[\hat{A}_1(k_x)\cos(\gamma_d z) - \hat{B}_1(k_x)\sin(\gamma_d z)\right] - \\ \mu\left(2\gamma_s^2 - k_s^2\right)\left[\hat{A}_2(k_x)\sin(\gamma_s z) + \hat{B}_2(k_x)\cos(\gamma_s z)\right]\end{aligned} \tag{20}$$

2.5 *Equations of Motion for the Fluid*

The following development assumes an inviscid and irrotational acoustic fluid with a homogeneous flow velocity, U. The pressure in the fluid with a bulk modulus of $\lambda_0 = \rho_0 c_0^2$ and signed Mach number, $M = U/c_0$, is obtained from the convected wave equation:

$$\left(1 - M^2\right)\frac{\partial^2 p(\mathbf{x})}{\partial x^2} + \frac{\partial^2 p(\mathbf{x})}{\partial z^2} - 2jMk_0\frac{\partial p(\mathbf{x})}{\partial x} + k_0^2 p(\mathbf{x}) = 0 \tag{21}$$

Substituting Eq. (21) into of Eq. (3a) results in the transformed convected wave equation for the acoustic pressure in the fluid:

$$\frac{\partial^2\hat{p}(k_x, z)}{\partial z^2} + \left[(k_0 - Mk_x)^2 - k_x^2\right]\hat{p}(k_x, z) = 0 \tag{22}$$

The solution to Eq. (22) is obtained by retaining only the outgoing waves (+z direction). It follows that:

$$\hat{p}(k_x, z) = \hat{C}(k_x)\,e^{-jk_z(z-h)} \tag{23}$$

where the real part of the exponent in Eq. (23) is restricted to negative values. It follows that:

$$k_z = \begin{cases} \left[(k_0 - Mk_x)^2 - k_x^2\right]^{1/2}, & (k_0 - Mk_x)^2 \geq k_x^2 \\ -j\left[k_x^2 - (k_0 - Mk_x)^2\right]^{1/2}, & (k_0 - Mk_x)^2 < k_x^2 \end{cases} \tag{24}$$

The z-component of transformed displacements in the fluid can be obtained from the following expression:

$$\hat{u}_z^f(k_x, z) = \frac{-jk_z\hat{C}(k_x)}{\lambda_0(k_0 - Mk_x)^2}e^{-jk_z(z-h)} \tag{25}$$

where the superscript f is used to denote a variable in the fluid volume V_f. The stress tensor for the acoustic fluid is diagonal and the transformed stresses and pressure can, respectively, be expressed as:

$$\hat{\tau}_{zz}^f(k_x, z) = -\hat{p}(k_x, z) \tag{26}$$

2.6 Coupled Equations

A coupled system of linear equations for the unknown constants can now be obtained by imposing boundary conditions on the viscoelastic solid at $z = 0$ and requiring compatibility of the stresses and displacements in the viscoelastic solid and fluid on the interface at $z = h$.

The transformed boundary conditions on $z = 0$ include zero transverse displacement and an imposed velocity distribution. It follows from Eqs. (17) and (18) that these boundary conditions are expressed as:

$$\hat{u}_x(k_x, z)\big|_{z=0} = 0 \tag{27}$$

$$j\omega\hat{u}_z(k_x, z)\big|_{z=0} = \hat{v}(k_x, z)\big|_{z=0} \tag{28}$$

where $\hat{v}(k_x, z)$ is the transformed normal velocity. For the case of a piston acting directly on the fluid, the spectral impedance is obtained by substituting Eqs. (25) and (26) into Eq. (5) resulting in:

$$\hat{Z}^f(k_x) = \rho_0 c_0\left(1 - M\frac{k_x}{k_0}\right)^2 / \sqrt{\left(1 - M\frac{k_x}{k_0}\right)^2 - \left(\frac{k_x}{k_0}\right)^2} \tag{29}$$

The transformed compatibility conditions on $z = h$ can be expressed as

$$\hat{u}_z^f(k_x, z)\Big|_{z=h} = \hat{u}_z(k_x, z)\big|_{z=h} \tag{30}$$

$$-j\omega\hat{u}_z^f(k_x, z)\Big|_{z=h}\hat{Z}^f(k_x) = \hat{\tau}_{zz}(k_x, z)\Big|_{z=h} \tag{31}$$

$$\hat{\tau}_{xz}(k_x, z)\Big|_{z=h} = 0 \tag{32}$$

The unknown $\hat{C}(k_x)$ in Eq. (26) can be expressed as a function of $\hat{A}_1(k_x)$, $\hat{B}_1(k_x)$, $\hat{A}_2(k_x)$, and $\hat{B}_2(k_x)$ by substituting Eqs. (18) and (25) into Eq. (30) and simplifying. Substituting Eqs. (17)–(20), and (30) into Eqs. (27), (28), (31), and (32) and simplifying results in the following set of linear equations for the complex unknowns:

$$\begin{bmatrix} t_{11} & 0 & 0 & t_{14} \\ t_{21} & t_{22} & t_{23} & t_{24} \\ 0 & t_{32} & t_{33} & 0 \\ t_{41} & t_{42} & t_{43} & t_{44} \end{bmatrix} \begin{Bmatrix} \hat{A}_1(k_x) \\ \hat{B}_1(k_x) \\ \hat{A}_2(k_x) \\ \hat{B}_2(k_x) \end{Bmatrix} = \begin{Bmatrix} \hat{v}(k_x) \\ 0 \\ 0 \\ 0 \end{Bmatrix} \tag{33}$$

where the t_{ij} are given in Appendix 1. The determinant of the coefficient matrix in Eq. (34) can be expressed as:

$$|\Delta| = [t_{21}t_{14} - t_{24}t_{11}][t_{43}t_{32} - t_{42}t_{33}] + [t_{23}t_{32} - t_{22}t_{33}][t_{44}t_{11} - t_{41}t_{14}] \tag{34}$$

The solution to Eq. (34) is obtained in closed form using standard methods. After some simplification, the transformed impedance in Eq. (5) can be expressed as:

$$\hat{Z}(k_x) = \rho\omega^2\gamma_s\left[\frac{t_{24}t_{41} - t_{21}t_{44}}{|\Delta|}\right] \tag{35}$$

The expression for impedance can now be found by substituting Eq. (35) into Eq. (9) with the following final result:

$$z = \frac{2\rho\omega^2}{\pi\delta^2}\int_{-\infty}^{\infty}\gamma_s\left[\frac{t_{24}t_{41} - t_{21}t_{44}}{|\Delta|}\right]\frac{\{\cos(k_xL/2) - \cos(k_xL/2 + k_x\delta)\}^2}{k_x^4}dk_x \tag{36}$$

The author is not aware of suitable closed form analytical or approximate methods for the general solution of Eq. (36). For the present study, the integral in Eq. (36) is solved using direct numerical integration.

3 Numerical Results and Discussion

Numerical results are presented in this section for the impedance of a piston acting on a viscoelastic solid that is under an inviscid homogeneous flow as shown in Fig. 1. Mechanical, physical, and geometric properties of the fluid and viscoelastic solid are listed in Table 1. The solid properties are representative of a typical polyurethane encapsulation material (CONAP EN-7 at 20 °C and 50 kHz). The frequency dependence of the viscoelastic solid mechanical properties is not accounted for in the present study. It is common for encapsulation materials to be chosen based on matching the plane dilatational wave characteristic impedance with the fluid. The ratio of characteristic plane wave impedances for the two layers is expressed as $R_m = |\rho c_d/\rho_0 c_0|$. For the present study, this ratio is calculated using the properties given in Table 1 as $R_m = 1.143$.

The spectral impedance of the moving fluid medium is presented first, followed by the spectral impedance of the two-layer medium including flow. Results for the impedance of a piston acting directly on a moving homogeneous medium are then presented and compared with the analytical solution. Next, results for the impedance of a piston on a two-layer medium with inviscid homogeneous flow are presented and discussed.

3.1 *Spectral Impedance*

The real and imaginary components of the normalized fluid spectral impedance are obtained from Eq. (29) and shown in Figs. 2 and 3, respectively, as a function of the normalized wavenumber. The region of acoustic radiation occurs at normalized wavenumbers from $1/(M-1)$ to $1/(M+1)$. Within this region, the fluid spectral impedance is real as indicated in Fig. 2. Outside this range, the spectral impedance is imaginary and positive indicating a mass-type reactance as shown in Fig. 3. The reactive spectral impedance of the moving medium has a zero that occurs for waves propagating in a direction opposite to the flow and at a normalized wavenumber of $1/M$. The reactive impedance for large wavenumber and $M = 0$ decreases as $1/k_x$. This is in contrast to the moving medium where the reactive impedance increases linearly with wavenumber.

Table 1 Material properties

Property	Value
ρ_0	1000 kg/m^3
c_0	1500 m/s
ρ	1050 kg/m^3
λ	2.44×10^9 N/m^2
μ	$1.80\times10^8(1+j0.150)$ N/m^2
L	0.005 m

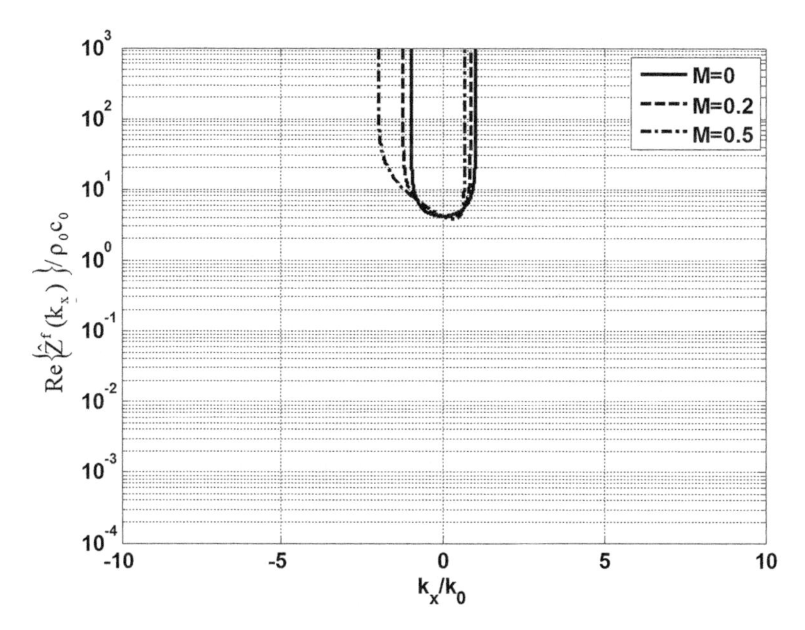

Fig. 2 Normalized real component of the fluid spectral impedance

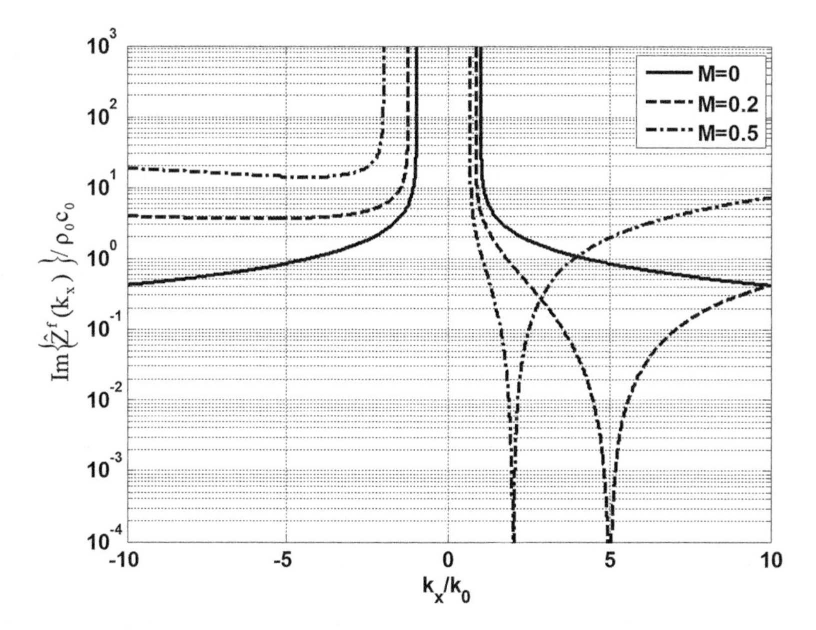

Fig. 3 Normalized imaginary component of the fluid spectral impedance

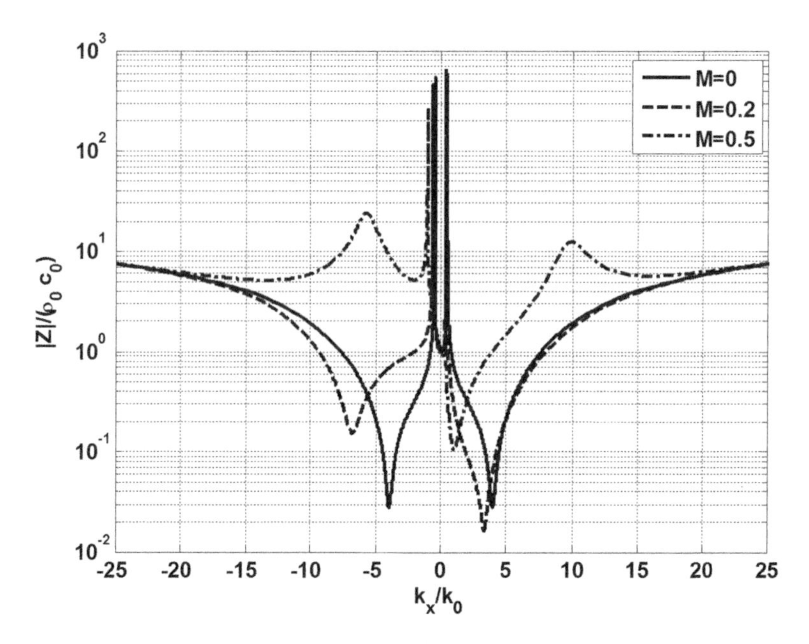

Fig. 4 Normalized spectral impedance magnitude for $h/L = 0.2$ and $k_0L = 0.5$ resistance

The magnitude of the normalized spectral impedance for the two-layer medium, obtained from Eq. (35) at a normalized frequency of $k_0L = 0.5$, is shown in Fig. 4. The spectral impedance at low and intermediate wavenumbers exhibits a strong Mach number dependence. In particular, the nulls are shifted in both amplitude and wavenumber. The local maxima occurs near the shifted acoustic wavenumbers at $1/(M-1)$ and $1/(M+1)$. In the high wavenumber region it is evident that the spectral impedances all asymptotically approach the value of a viscoelastic half-space. This asymptotic behavior is a result of the viscoelastic solid containing only evanescent waves when $k_x > k_d$ and $k_x > k_s$. Furthermore, the spectral impedance increases linearly with k_x in the high wavenumber region. This behavior is important when considering the convergence of the integral in Eqs. (9) and (36).

3.2 Impedance of Piston Acting Directly on Moving Medium

The radiation impedance of a piston acting directly on a moving medium is shown in Fig. 5. This result is a special case where Eq. (10) is solved using direct numerical integration with the fluid spectral impedance obtained from Eq. (29). The numerical results are noted to be in excellent agreement with the analytical solution [10] for the real component of the impedance. As the Mach number is increased, the peak in the radiation resistance is shifted to lower frequencies. It is further noted that the imaginary component of radiation impedance obtained from Eq. (10) does not

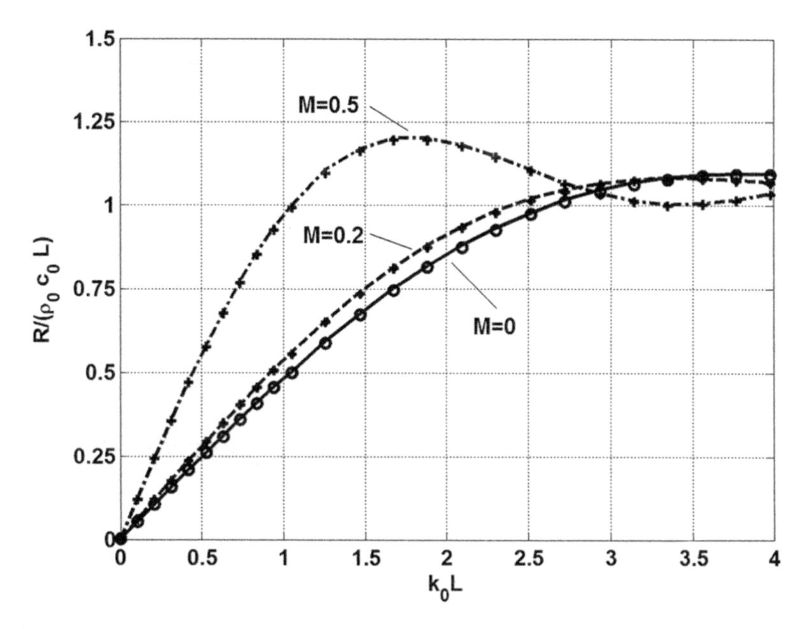

Fig. 5 Radiation resistance of piston acting directly on a moving medium. Direct numerical integration (lines) and analytical solution [10] (symbols)

converge for $M > 0$ as a result of the integration kernel rolling off as $1/k_x$. This can be attributed to the boundary condition on the flow above the piston and rigid baffle that results in a Dirac delta functions at $x = \pm L/2$. As a result of the flow, the upstream edge gives rise to a line source and the downstream edge gives rise to a line sink.

3.3 Impedance of Piston on a Two-Layer Medium with Homogeneous Flow

Real and imaginary components of the piston impedance with viscoelastic layer thickness $h/L = 0.2$ and aperture function parameter $\delta/L = 0.02$ for various Mach numbers are shown in Figs. 6 and 7, respectively. Also shown are classical results for the case with $M = 0$ and $h/L = 0$ which corresponds to the case of a piston acting directly on a non-moving fluid. From these results it is clearly evident that the moving fluid medium significantly influences the resistive and reactive components of the impedance at the lower frequencies. In particular, the results show that the low frequency reactive impedance shifts from negative (mass-like) to positive (spring-like) as the Mach number is increased. Additionally, the $M = 0.5$ case exhibits a significant deviation from the lower flow velocities. The low frequency reactive impedance at $M = 0$ is due to the shear stiffness of the viscoelastic medium and the

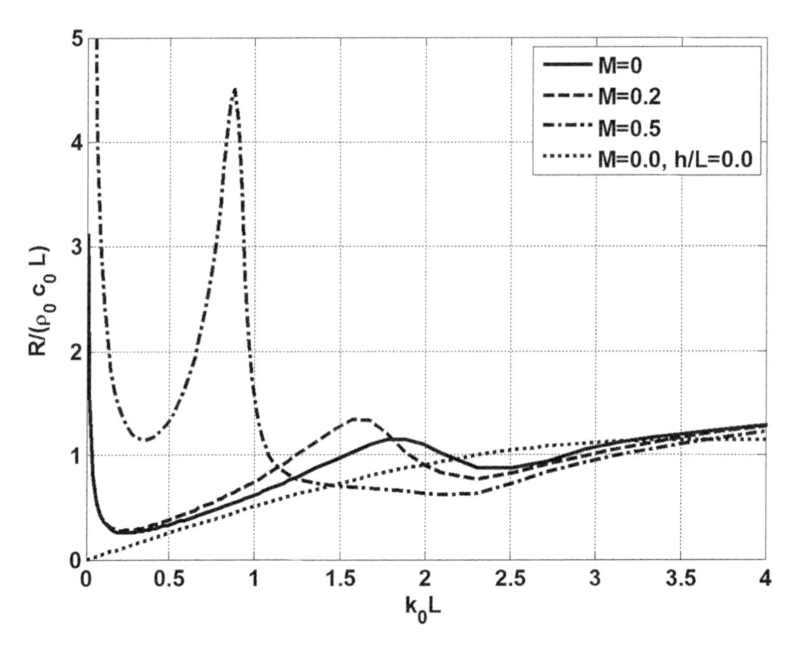

Fig. 6 Normalized resistance of a piston with h/L = 0.2 and δ/L = 0.02

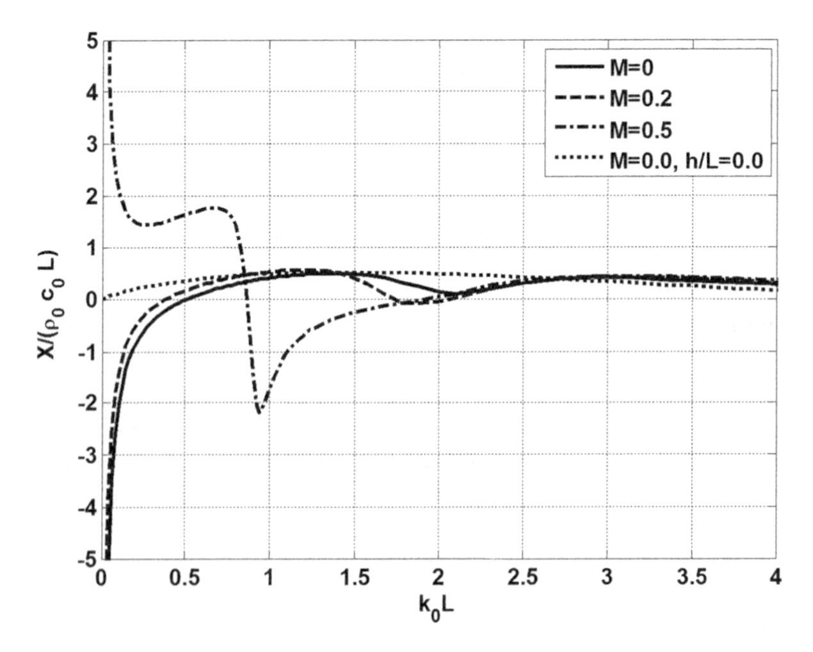

Fig. 7 Normalized reactance of a piston with h/L = 0.2 and δ/L = 0.02

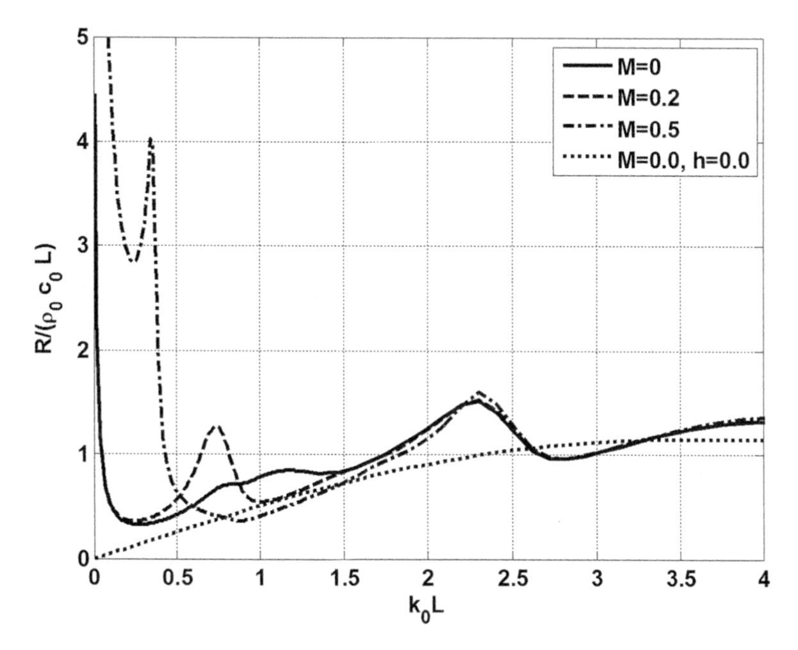

Fig. 8 Normalized resistance of a piston with h/L = 0.5 and δ/L = 0.02

resistance is due to the associated shear loss factor. This also occurs for the case of a moving medium, but is limited to the lower Mach numbers. The high frequency results approach but are not equal to the classical piston. The differences being due to the complex shear modulus and loss factor associated with the viscoelastic layer.

A similar set of results for the case with $h/L = 0.5$ and aperture function parameter $\delta/L = 0.02$ for various Mach numbers are shown in Figs. 8 and 9. In a qualitative sense, these results exhibit a behavior similar to the $h/L = 0.2$ results. The increase in layer thickness results in the first peak in the resistance to occur at a lower frequency. This is clearly evident in Figs. 10 and 11 that exhibit the influence of layer thickness at $M = 0.5$. It is also clear that the low frequency shifting from a stiffness reactance to a mass reactance is a function of both the layer thickness and flow velocity, but does not occur for a zero flow velocity.

Collectively, the results presented in this section illustrate the significant influence of both flow velocity and viscoelastic layer thickness on the resistance and reactance of a baffled piston. These results are of relevance to both the transmission and reception of acoustic waves in a moving medium. The design of electro-acoustic transduction systems requires accurate estimates of the piston radiation impedance. From a sound transmission perspective, the large variations in impedance would influence impedance matching and the typical goal of maximum acoustic power transfer from piston to the moving fluid medium. This effect is clearly evident in Figs. 6–11 which exhibits a strong dependence of impedance on frequency, Mach number, and layer thickness. When acting as part of an acoustic receiver system, it is

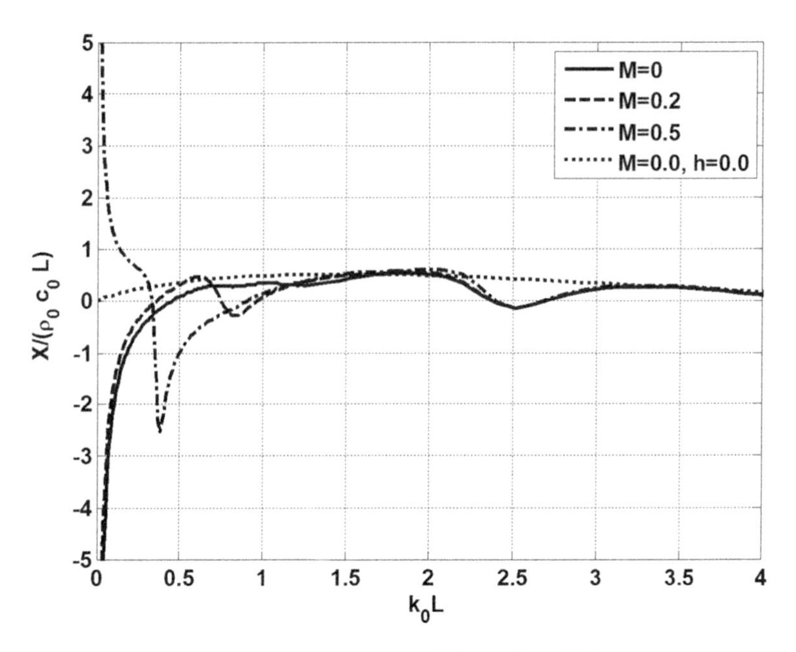

Fig. 9 Normalized reactance of a piston with h/L = 0.5 and δ/L = 0.02

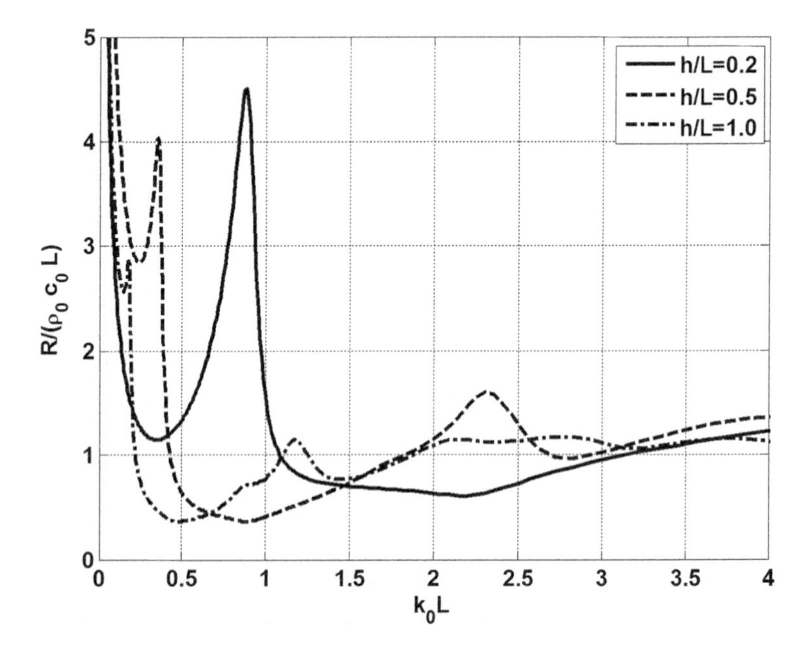

Fig. 10 Normalized resistance of a piston with M = 0.5 and δ/L = 0.02

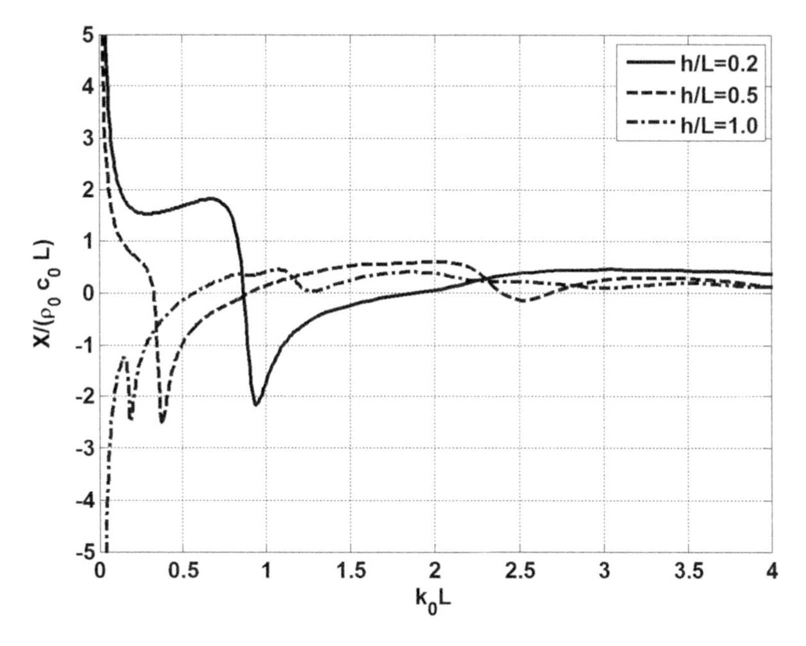

Fig. 11 Normalized reactance of a piston with M = 0.5 and δ/L = 0.02

clear from the asymmetry in wavenumber response shown in Fig. 4 that the upstream and downstream response of any transducer would be significantly influence by the Mach number in addition to layer thickness. Furthermore, the high wavenumber (non-propagating) nulls shown in Fig. 4 may also have significance to the design of devices and structures for self-noise suppression.

4 Summary

An approach to solve for the impedance of a piston in a rigid baffle acting on a two-layer semi-infinite medium with flow has been developed. The basic approach utilizes integral transform methods applied to the differential equations for the displacement potentials in the viscoelastic solid and convected wave equation for the fluid pressures. The general results are compared with limiting cases where closed form solutions are available. The presented results clearly show that the influence of Mach number and viscoelastic layer thickness on the resistive and reactive components of the impedance is most dominant in the low-mid frequency range. This basic approach can be readily extended to both circular and rectangular pistons.

Acknowledgments This work was supported by funds from the In-house Laboratory Independent Research Program at the Naval Undersea Warfare Center Division, Newport, Rhode Island.

Appendix 1

Elements of the coefficient matrix in Eq. (34):

$$
\begin{aligned}
t_{11} &= j\omega\gamma_d, \\
t_{12} &= 0, \\
t_{13} &= 0, \\
t_{14} &= -\omega k_x,
\end{aligned}
$$

$$
\begin{aligned}
t_{21} &= -\left(\lambda k_d^2 + 2\mu\gamma_d^2\right)\sin\left(\gamma_d h\right) + \tfrac{j\omega\gamma_d}{k_0}\hat{Z}^f\cos\left(\gamma_d h\right), \\
t_{22} &= -\left(\lambda k_d^2 + 2\mu\gamma_d^2\right)\cos\left(\gamma_d h\right) - \tfrac{j\omega\gamma_d}{k_0}\hat{Z}^f\sin\left(\gamma_d h\right), \\
t_{23} &= 2\mu j k_x\gamma_s\cos\left(\gamma_s h\right) - \tfrac{\omega k_x}{k_0}\hat{Z}^f\sin\left(\gamma_s h\right), \\
t_{24} &= -2\mu j k_x\gamma_s\sin\left(\gamma_s h\right) - \tfrac{\omega k_x}{k_0}\hat{Z}^f\cos\left(\gamma_s h\right),
\end{aligned}
$$

$$
\begin{aligned}
t_{31} &= 0, \\
t_{32} &= j k_x, \\
t_{33} &= -\gamma_s, \\
t_{34} &= 0,
\end{aligned}
$$

$$
\begin{aligned}
t_{41} &= 2\mu j k_x\gamma_d\cos\left(\gamma_d h\right), \\
t_{42} &= -2\mu j k_x\gamma_d\sin\left(\gamma_d h\right), \\
t_{43} &= \mu\left(2\gamma_s^2 - k_s^2\right)\sin\left(\gamma_s h\right), \\
t_{44} &= \mu\left(2\gamma_s^2 - k_s^2\right)\cos\left(\gamma_s h\right).
\end{aligned}
$$

References

1. L. Kinsler, A. Frey, *Fundamentals of Acoustics* (John Wiley & Sons, New York, 1964), pp. 176–182
2. M. Junger, D. Feit, *Sound, Structures, and Their Interaction* (AIP, New York, 1993), pp. 92–139
3. P. Stepanishen, The time dependent force and radiation impedance on a piston in a rigid infinite planar baffle. J. Acoust. Soc. Am. **49**, 841–849 (1971)
4. A. Pierce, R. Cleveland, M. Zampolli, Radiation impedance matrices for rectangular interfaces within rigid baffles: Calculation methodology and applications. J. Acoust. Soc. Am. **111**, 672–684 (2002)
5. J. Luco, Impedance functions for a rigid foundation on a layered medium. Nucl. Eng. Des. **31**, 204–217 (1974)
6. J. Luco, Vibrations of a rigid disc on a layered viscoelastic medium. Nucl. Eng. Des. **36**, 325–340 (1976)

7. S. Hassan, Impedance of pistons on a two-layer medium in a planar infinite rigid baffle. J. Acoust. Soc. Am. **122**, 237–246 (2007)
8. J.E. Ffowcs Williams, D.J. Lovely, Sound radiation into uniformly flowing fluid by compact surface vibration. J. Fluid Mech. **71**, 689–700 (1975)
9. F. Leppington, H. Levine, The effect of flow on the piston problem of acoustics. J. Sound Vib. **62**, 3–17 (1979)
10. H. Levine, A note on sound radiation into a uniformly flowing medium. J. Sound Vib. **71**, 1–8 (1980)
11. Y.M. Chang, P. Leehey, Acoustic impedance of rectangular panels. J. Sound Vib. **64**, 243–256 (1979)
12. D. Li, P. Stepanishen, Transient and harmonic fluid loading on vibrating plates in a uniform flow field via a wave-vector/time domain method. J. Acoust. Soc. Am. **83**, 474–482 (1988)
13. P. Morse, K. Ingard, *Theoretical Acoustics* (McGraw-Hill, New York, 1968), pp. 392–394
14. J. Naze Tjotta, S. Tjotta, Nearfield and farfield of pulsed acoustic radiators. J. Acoust. Soc. Am. **71**, 824–834 (1982)
15. J. Achenbach, *Wave Propagation in Elastic Solids* (North-Holland, New York, 1973)

Acoustics of a Mixed Porosity Felt Airfoil

Aren M. Hellum

1 Introduction

Lifting surfaces such as wings, hydrofoils, and propulsor blades produce noise in operation. This noise has several potential sources, including stall, shed vortices, and trailing edge scattering [1, 2]. Although the porous wings of owls have been shown [3, 4] to reduce the noise produced in flight, it is not clear at present what mechanisms are operating to produce this noise reduction. It is therefore worth investigating the form of solution favored by owls, the "mixed porosity" wing, in which some fraction of the lifting area is porous and the rest is impermeable.

The potential acoustic benefits of porous foils can be estimated from existing literature. Geyer et al. [5] measured sound reduction of 5–15 dB for airfoils made entirely of porous material. A 1973 patent awarded to Hayden and Chanaud [6] describes measurements using a porous foil in air, finding an 8–18 dB improvement over a reference solid foil. That patent also describes a number of mixed porosity arrangements, but does not specify a preferred arrangement. The patent literature on this topic is extensive [7–9], but typically lacks quantitative claims. The literature also contains other sources pertaining to porous lifting surfaces which are concerned primarily with cooling high-temperature turbine blades [10] or mitigating shock waves [11, 12]. Theoretical work on a poroelastic boundary layer [13] found less effect than the experimental works, finding a maximum improvement of 6 dB for a rigid, porous edge and up to 12 dB for a poroelastic edge.

A. M. Hellum (✉)
Naval Undersea Warfare Center, Newport, RI, USA
e-mail: aren.hellum@navy.mil

A. A. Ruffa, B. Toni (eds.), *Advanced Research in Naval Engineering*, STEAM-H: Science, Technology, Engineering, Agriculture, Mathematics & Health, https://doi.org/10.1007/978-3-319-95117-1_2

Porous materials can be parameterized in several ways. The inviscid model developed in [13] incorporates the bending wavenumber and density ratio between the fluid and foil. This model indicated that the flexible trailing edge of an owl's wing was responsible for most of that animal's ability to reduce trailing edge noise. Hayden and Chanaud [6] use the specific acoustic impedance of the porous material relative to the surrounding medium. The pressure drop of air through a unit thickness of the material (resistivity) has also been used to describe the porous material [5]. Resistivity is fluid- as well as material-dependent. Various "low-level" porous material identifiers such as the nominal size of pores, fibers, and particles have also been employed [14, 15]. These parameterizations are not readily comparable, so the present work uses a strong-performing material from previous studies in order to examine mixed porosity arrangements.

The lack of consensus regarding the parameterization of porous lifting surfaces reflects a similar confusion about the mechanisms at work, but it is generally held that porous materials weaken the "noise scattering" mechanism at the trailing edge. As the boundary layer convects past the trailing edge, the pressure field goes from being supported by the foil surface to being unsupported in the wake, "scattering" half of its energy in the process [8]. The porous area fraction and porous material parameters which can beneficially interact with the noise scattering mechanism are not clear based on the present literature.

In this work, measurements taken for three porous area fractions are presented. The test apparatus was designed to preserve the elastic qualities of the foil while varying the porous area fraction, in order to isolate the effect of the latter variable. The noise reduction produced by the fully porous arrangement is shown to match the most applicable published data. The noise reduction produced by the mixed porosity arrangement is found to be similar to the fully porous arrangement at low Reynolds numbers over the frequency range measured. The noise reduction produced by the mixed and fully porous foils does not match over the entire frequency range at higher Reynolds numbers. These trends in noise reduction and possible explanations for the effect are discussed.

2 Results

The present data were acquired in the 48″ acoustic wind tunnel (Fig. 1a) at the Naval Undersea Warfare Center in Newport, Rhode Island (NUWC). The test item used was made of dense felt, laser cut into a NACA0012 airfoil shape with an 11.5″ chord. The porous section's 24″ span was constructed from ¾″ sections (Fig. 1b). The alignment shafts were used to support the foil during the test. The porous section is shown in 100% and 54% porous configurations in Fig. 1c, d. The remainder of the 53″ span is made of 3D printed ABS plastic (blue in Fig. 1a, c and d); this span ensures that the airfoil tips are not in the flow. Additional details regarding the setup, data acquisition, and processing are provided in Sect. 5.

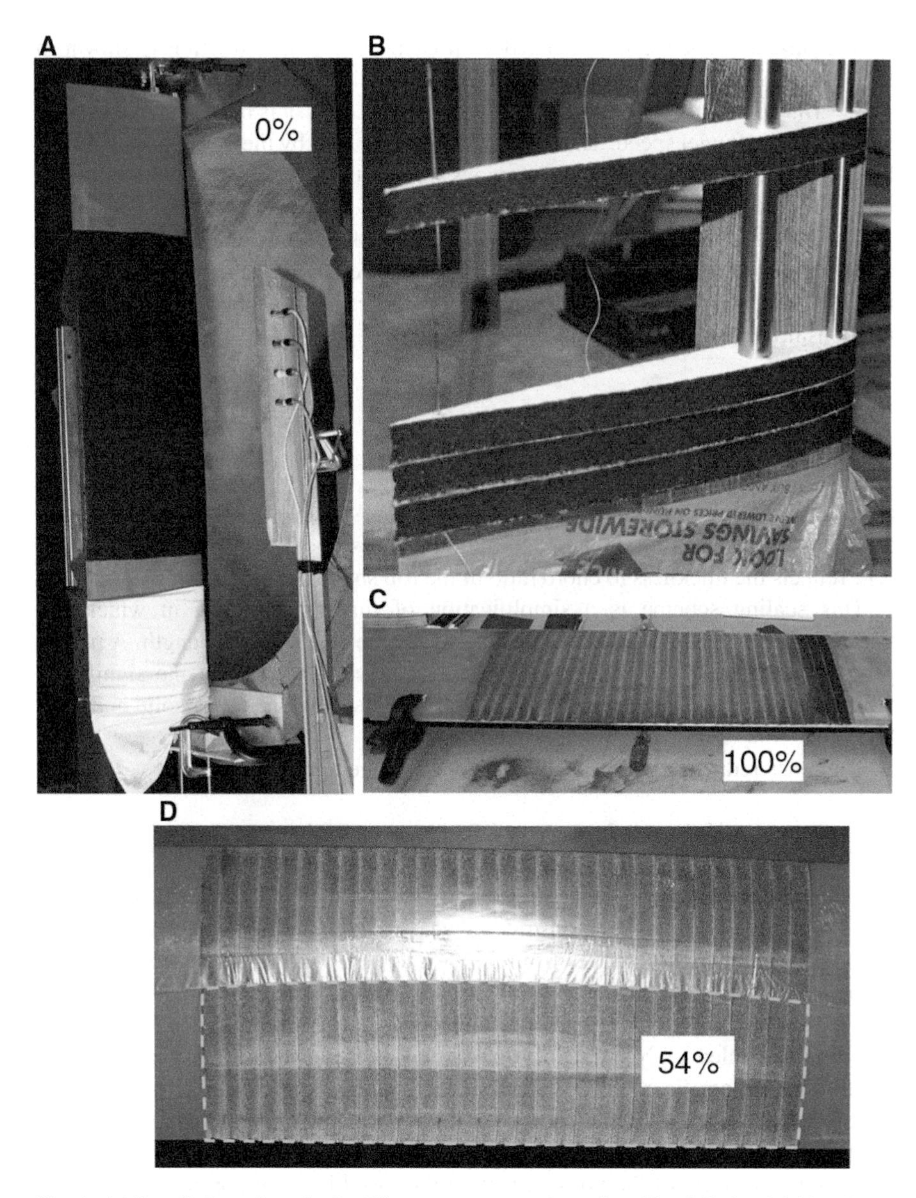

Fig. 1 (**a**) Installed test item in the 0% porous area configuration. The full-chord fairing is the central black section. The four-element microphone array is at right. A spandex cover (white) which was stretched over the length of the test item during all tests has been removed and is at the bottom of the foil. (**b**) Felt section during construction; alignment shafts are at right of image. (**c**) Test item in 100% porous area fraction configuration during construction. The darker color at right is unremoved char layer (Sect. 5). (**d**) Test item in 54% porous area fraction configuration. The porous trailing edge is outlined

The unscaled spectrum associated with each porous area fraction is shown in Fig. 2 over a range of Reynolds numbers. These data were taken over the range $U_\infty = 18.4$–30.5 m/s. The chord Reynolds number $U_\infty L_c/\nu$ is indicated in each plot to indicate the tunnel speed. The elevated noise produced by the 100% porous foil at high frequencies has been noted in previous studies; this phenomenon is investigated further in Sect. 3.

These data also indicate that—as expected—the frequency and amplitude of the sound produced by the foils are functions of the free-stream velocity. A scaled and unscaled version of the baseline (0% porosity) curves is shown in Fig. 3a, where the relationships [16]

$$SPL_{scaled} = SPL - 10\log_{10} U_\infty^5 \text{ dB}, \quad St_d = \frac{f(0.12c)}{U_\infty}$$

are used to scale the ordinate [16] and abscissa, respectively. The latter scaling is a Strouhal number based on the maximum thickness of the foil where the constant 0.12 reflects the thickness to chord ratio of the foil shape being tested (NACA0012).

This scaling scheme is a simplification of that used in [1], in which the trailing-edge boundary layer thickness is used as the characteristic length. A purely geometric scheme was preferred for the present work because of the significant difference between the trailing edge boundary layers developed over solid and foils (Fig. 3b).

The noise reduction produced by a porous arrangement is defined as

$$Noise\ reduction = SPL_{porous} - SPL_{nonporous}$$

The noise reduction for each foil is shown in Fig. 4. The noise reduction of the 100% porous foil (Fig. 4a) demonstrates a weak trend in Reynolds number whereby the peak noise reduction moves from $St_d \approx 6$ to $St_d \approx 4.5$ as the speed increases. Each tested speed reduces noise by at least 9 dB over at least one-third octave band, with higher values of noise reduction at higher speeds. The trend for the 54% porous foil (Fig. 4b) is cleaner. Noise reduction is largest at the lowest speeds and decreases to a minimum near $Re_c \approx 5 \times 10^5$ before rebounding slightly. It is possible that if each foil were tested at higher speeds, this rebound would continue, but this could not be tested because of speed limitations imposed by the foil construction (Sect. 5).

An alternate view of the noise reduction produced by each porous arrangement is provided in Fig. 5, in which the two arrangements are compared directly. It is interesting that at the lowest Reynolds numbers, the noise reduction of the 54% porous foil essentially tracks that of the 100% porous foil over the entire frequency range. At higher speeds, the two porous arrangements clearly diverge. This transition is investigated further in Sect. 3.

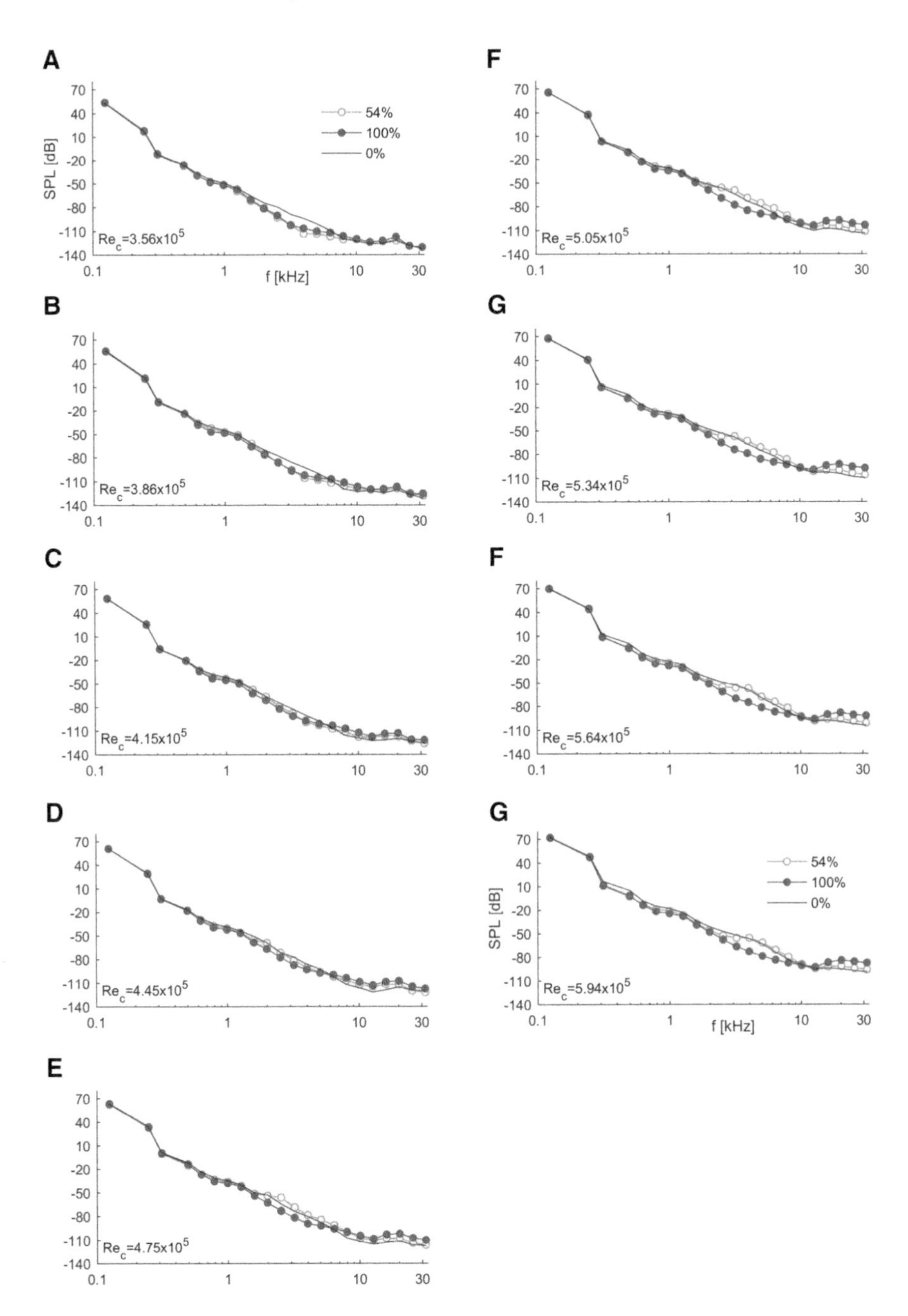

Fig. 2 Measured spectra for 0% porous foil, 54% porous foil, and 100% porous foil

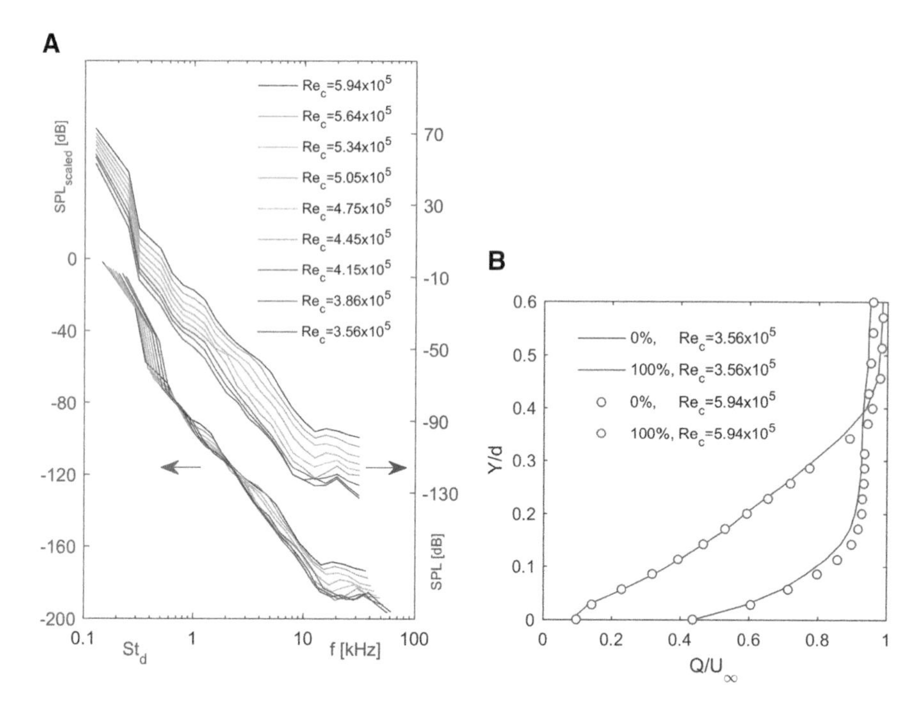

Fig. 3 (**a**) Comparison of scaled and unscaled coordinates, 0% porous baseline foil. Left axes: Scaled sound pressure levels vs. thickness-based Strouhal number St_d. Right axes: Unscaled sound pressure levels vs. frequency, in kHz. (**b**) Wake at 3.2 mm downstream of trailing edge, obtained via hotwire anemometry. These surveys were acquired behind 100% porous and 0% porous sections at the highest and lowest Re_c used in the present test. Further details of this survey are provided in Sect. 5

3 Discussion

3.1 Comparison with Published Results

The present data are compared with the previously published data of Geyer et al. [5] and Herr et al. [14] in Fig. 6. These studies employed a similar wool felt in their material survey. Relevant parameters of each study are provided in Table 1. The published studies tested multiple porous materials at multiple speeds; only the combination being compared is listed. The published works used a separate rigid airfoil or insert as an impermeable baseline while the present data covered the felt foil with a thin plastic membrane to approximately preserve the elastic qualities of the foil.

The similarity between the two 100% porous curves is striking in light of the differences in apparatus employed in the present and the Geyer et al. studies.

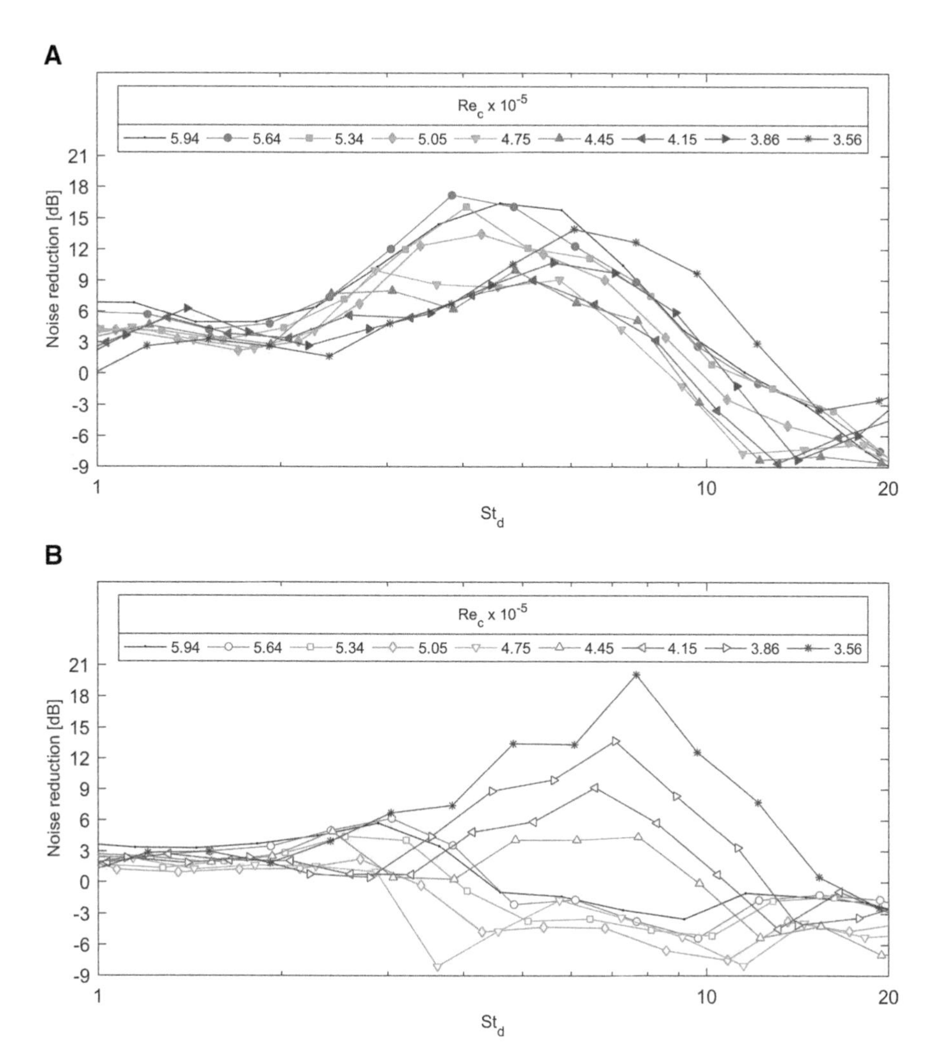

Fig. 4 Noise reduction produced by porous foils relative to non-porous foil. (**a**) 100% porous area fraction. (**b**) 54% porous area fraction

Because the present study used a non-rigid impermeable baseline, this similarity indicates that the elasticity of the foil is a secondary factor contributing to noise reduction [13]. The noise reduction measured for the 54% porous foil is more similar to that measured by Herr at 10% porous area fraction than it is to the 100% porous foil from the present study. The fact that the partially porous foils show similar levels of noise reduction at very different porous area fractions is interesting, and a refined version of the present study is planned to investigate this further.

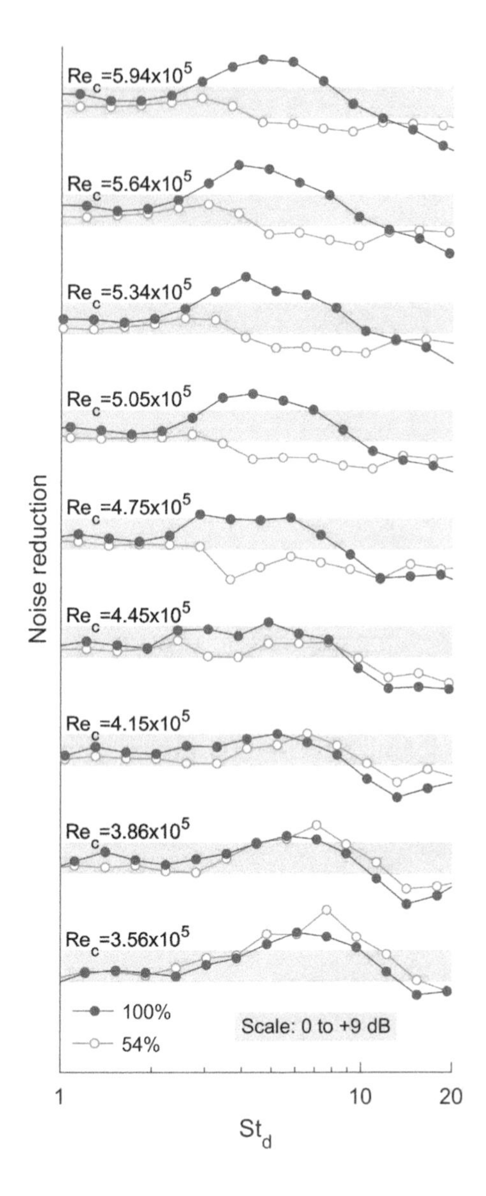

Fig. 5 Noise reduction produced by each porous arrangement at all tested Re_c. The height of the shaded region indicates 0 to +9 dB of noise reduction for each set of curves

3.2 Elevated High-Frequency Noise, Physical Model

Each study indicates that increased noise is produced by porous foils at high frequencies; data reflecting this do not appear in Geyer's figures but it is indicated in that work's text. One theory proposed to explain this phenomenon [5, 14] attributes this elevated noise to the difference in surface roughness between the porous and reference surfaces. This explanation is unsatisfactory based on the present study,

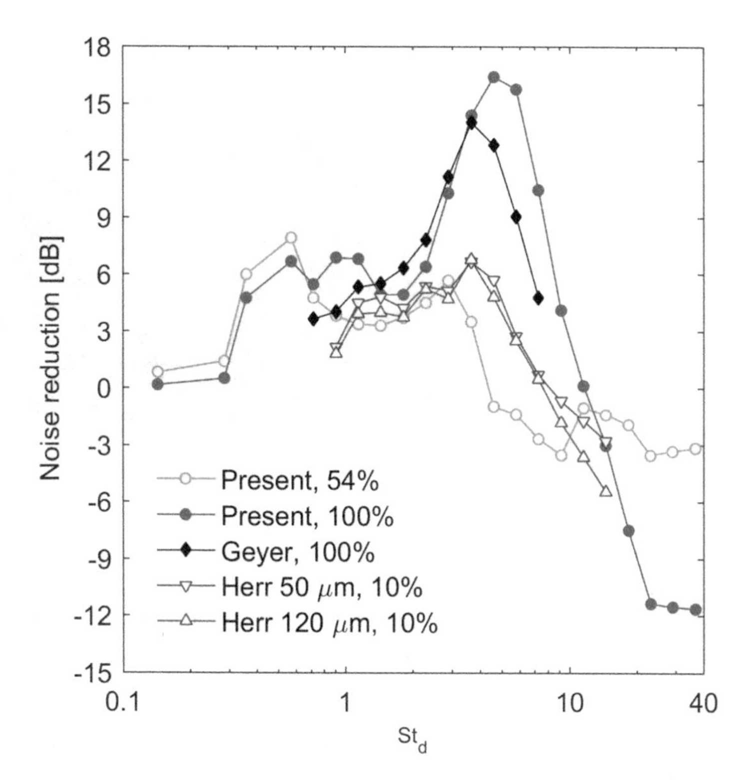

Fig. 6 Noise reduction data from the present study, Geyer et al. [5], and Herr et al. [14]. Present data were collected at $Re_c = 5.94 \times 10^5$. The percentage indicates porous area fraction and the length measurement indicates the pore size

Table 1 Relevant parameters of the present study and published porous airfoil studies using similar materials

Study	Porous material	Shape	Thickness (mm)	Area fraction	Re_c	Array
Present	Dense felt	NACA0012	35	100%, 54%	5.9×10^5	Line, 4 element
Geyer [5]	Dense felt	SD7003	20	100%	1.0×10^6	Planar, 56 element
Herr [14]	Sintered fiber felt	"DLR-F16 model"	37	10%	1.3×10^6	Planar, 96 element

wherein a fabric cover was placed over the foil to preserve the surface roughness between porosity conditions (Sect. 5).

Because the elasticity and the surface roughness of the foil was conserved between porous and non-porous runs, the following argument based on percolation through the porous medium is proposed. The pressure field characteristic of a fluid convecting past a non-porous surface is fully supported by the surface. This is commonly described in terms of "image" vortices mirroring the flow field inside the

Fig. 7 Scanning electron microscope image of wool felt used in the present study

solid wall [17]. It is this pressure support which is scattered acoustically, leading to the amplification of sound at the trailing edge [18, 19]. In contrast, pressure variations near a porous surface induce flows within the medium which percolate through the porous surface. These percolating flows are subjected to breakup by the porous medium, producing high frequency sound.

Features of the porous medium and the boundary layer can be used to compute a characteristic frequency of this high frequency sound. At $Re_c = 5.94 \times 10^5$, flat-plate turbulent boundary layer relationships [20] indicate a wall shear stress of $\tau_w = 2.16$ Pa at the trailing edge. This yields a friction velocity $u_\tau = \sqrt{\tau_w/\rho} = 1.34$ m/s, which characterizes the velocities in the viscous sublayer region of the boundary layer, nearest the foil's surface. Per Fig. 7 and similar work in [14], we choose a nominal length scale for the pores to be $\ell_{pore} \approx 100$ µm. These scales yield a characteristic percolation frequency $f_{perc} = u_\tau \ell_{pore} = 13.4$ kHz, or $St_d = 15.4$ at $Re_c = 5.94 \times 10^5$. This is a reasonable order-of-magnitude estimate of the frequency at which the porous foil begins to produce increased noise, visible in Figs. 4 and 6.

3.3 Noise Reduction Trend with Reynolds Number

The noise-reducing qualities of the 100% porous and 54% porous foils do not react the same way to increasing Reynolds number. This is demonstrated clearly in Figs. 4 and 5. The Reynolds number range used in this study indicates the possible role of turbulent transition [21]. Arguing against this hypothesis is that some care was taken to trip the flow (Sect. 5), and tufting was performed which indicated that the foil was not experiencing laminar separation [22]. It may be that despite the scaling relevance of trailing edge flow scales on acoustics [1], leading edge porosity is more important to the development of a "quiet" boundary layer. A refined version of the present study is planned to investigate this point.

4 Concluding Remarks

The present data indicate that a foil with 100% porous chord fraction is capable of reducing noise over a significant band of frequencies over the tested velocity range. The same foil, faired to produce a 54% porous chord fraction, performs similarly at low speeds, but is less effective at higher Reynolds numbers.

When the elasticity of the trailing edge is preserved between impermeable and porous foils, substantial noise reduction is still realized, indicating that the role of this property is likely to be secondary.

Increased high frequency noise is produced by porous foils despite the care taken to preserve the boundary layer roughness. A plausible physical model for this increased high frequency noise based on fluid percolation through the porous medium is presented.

5 Methods

5.1 Data Acquisition and Processing

A line array of four Bruel and Kjaer microphones (Fig. 1a) was used to acquire the acoustic data at 60 kHz over 30 s. Spacing between the microphones was 2″(5.1 cm). This line array was affixed parallel to the trailing edge. The array was placed as close as possible to the trailing edge in the chord-normal direction without obstructing the developing shear layer produced by the jet. This distance is just over 24″ (61 cm). The center of this array corresponded with the center of the porous span of the foil. No beamforming was used to separate the noise produced by the porous sections of the span from that produced by the impermeable sections.

A background noise file was acquired at the beginning and end of each data session with the wind tunnel turned off. The average of these background spectra was subtracted from each data file. The spectra were produced using MATLAB's pwelch (Power Spectral Density, Welch's method) routine. The averaging window was a 512 point Hanning window with 50% overlap. The spectral data presented here have been averaged over one-third octave bands, with $f_{ctr} = 1$ kHz.

5.2 Foil Construction

The 11.5″ chord foil cross sections were produced using a laser cutter. Each spanwise slice is ¾″ thick. The cutting procedure produces a burned layer on the surface of the foil which is removed by sanding. This dark, burned layer covers the surface of each foil slice in Fig. 1b and is shown partially removed in Fig. 1c, where the darker section on the far right side of the porous section is unremoved in the latter image. The thickness of the felt trailing edge could not be built to the specified fineness of the NACA0012 shape following the laser cutting and sanding procedures. A thickness of ≈ 1 mm with some spanwise variation was measured, comparable to that obtained in prior studies [5].

A Kevlar thread was passed through the felt cross-sections 1″ upstream of the trailing edge. It was affixed to a tension-producing windlass, for the purpose of strengthening the trailing edge at high angles of attack. Figure 1b shows this thread being put in place during the construction of the porous section. No additional strength was found to be necessary during trials and the thread was removed prior to data collection.

A Spandex cover was placed over the faired foil in order to preserve the surface roughness between tests. This cover is the white fabric at the bottom of the span in Fig. 1a, having been removed to show the faired section of span. The seam of this cover was placed at the leading edge of the foil to introduce a mild corrugation at the leading edge. It was found that running with the cover in place removed the aeroacoustic resonance that was apparent at some speeds. This is taken as an indication that the boundary layer was sufficiently tripped at the trailing edge of the foil. Originally, a Kevlar thread was stretched over the entire span of the foil just ahead of the ¼-chord position as a trip. This was found to yield similar results, but some combinations of flow speed and tension would produce a resonant buzz which polluted the acoustic data. The maximum speeds used in the test were constrained by the tendency of this cover to flap.

The fairing was made of thin polyethylene, black and clear in Fig. 1a, d, respectively. After some experimentation, standard home garbage bags were found to be sufficient for this purpose. In the 0% porous configuration, the fairing was stretched taut around the foil. The free edge was secured with packing tape on the side of the foil opposite from the microphone array. A heat gun was then used to shrink the fairing around the foil to tighten it, with care taken not to create wrinkles in the surface. At high speeds, the fairing has a tendency to "puff" slightly because of aerodynamic loading, separating it from the felt surface. A stronger material would reduce this tendency and may allow the test to be run at higher speeds. For the present test, a thin fairing was selected in order to better preserve the elastic qualities of the trailing edge.

In the 54% porous configuration, the free edges of the fairing could not be firmly attached to the felt surface of the foil, so an alternate method was used. A seam was created in the free edge of the fairing and a Kevlar thread was run through this seam. This thread was placed under tension and secured. The heat gun was then used to shrink the fairing as in the 0% porous case. This tensioning creates the curved rear edge of the fairing which is visible in Fig. 1d. Stitches were then placed through the Spandex cover, fairing, and airfoil with a tapestry needle to tack the Kevlar thread to the surface. Three evenly spaced stitches in the spanwise direction were found to be sufficient to keep the fairing from flapping. A fairing condition, once removed from the foil, could not be reliably repeated. Because the foil elasticity was not found to substantially affect the foil's acoustics, an improved apparatus has been designed for future tests.

5.3 *Hotwire Anemometry*

Hotwire anemometry was used to conduct a survey of the wake immediately behind the solid and porous sections of the foil's span, at a distance of 3.2 mm. The direction of survey was in the chord-normal direction, taken with a spacing of 1 mm. A point was also acquired at a distance of 100 mm from the chord to measure the free-stream velocity. Each data point is an average of 5 s of data acquired at 1 kHz. The survey was conducted using a single-wire DANTEC probe, with its axis oriented parallel to the trailing edge. Use of this probe means that the velocity being measured is a combination of the streamwise (u) and chord-normal components (v) of velocity, yielding the "cooling velocity" $Q = \sqrt{u^2 + v^2}$.

References

1. T. Brooks, D. Pope, M. Marcolini, Airfoil self-noise and prediction, NASA, vol. 1218, 1989
2. R. Amiet, Noise due to turbulent flow past a trailing edge. J. Sound Vib. **47**(3), 387–393 (1976)
3. T. Bachmann, S. Klän, W. Baumgartner, M. Klaas, W. Schröder, H. Wagner, Morphometric characterisation of wing feathers of the barn owl Tyto alba pratincola and the pigeon Columba livia. Front. Zool. **4**(1), 23 (2007)
4. E. Sarradj, C. Fritzsche, T. Geyer, Silent owl flight: bird flyover noise measurements. AIAA J. **49**(4), 769–779 (2011)
5. T. Geyer, E. Sarradj, C. Fritzsche, Measurement of the noise generation at the trailing edge of porous airfoils. Exp. Fluids **48**, 291–308 (2010)
6. R. Hayden, R. Chanaud, Method of reducing sound generation in fluid flow systems embodying foil structures and the like. U.S. Patent 3,779,338, 1973
7. C. Ellett, Noiseless propeller. U.S. Patent 2,340,417, 1 Feb 1944
8. A. Gupta, T. Maeder, Wind-turbine blade and method for reducing noise in wind turbine. U.S. Patent 7,901,189, 2011
9. P. Ho, P. Gliebe, Low noise permeable airfoil. U.S. Patent 6,139,259, 2000
10. D. Kump, N. Lauziere, Method of controlling the permeability of a porous material, and turbine blade formed thereby. U.S. Patent 3,402,914, 24 Sept 1968
11. P. Hartwich, Porous airfoil and process. U.S. Patent 5,167,387, 1 Dec 1992
12. R. Mineck, P. Hartwich, Effect of full-chord porosity on aerodynamic characteristics of the NACA 0012 airfoil, NASA, 1996
13. J. Jaworski, N. Peake, Aerodynamic noise from a poroelastic edge with implications for the silent flight of owls. J. Fluid Mech. **723**, 456–479 (2013)
14. M. Herr, K. S. Rossignol, J. Delfs, M. Mößner, N. Lippitz, Specification of porous materials for low-noise trailing-edge applications, in *20th AIAA/CEAS Aeroacoustics Conference*, Atlanta, GA, Paper 3041-2014
15. M. Herr, A noise reduction study on flow-permeable trailing-edges, Deutsches Zentrum für Luft-und Raumfahrt, Institute of Aerodynamics and Flow Technology, Braunschweig, 2007
16. T. Brooks, T. Hodgson, Trailing edge noise prediction from measured surface pressures. J. Sound Vib. **78**, 69–117 (1981)

17. J. Katz, A. Plotkin, *Low-Speed Aerodynamics* (Cambridge University Press, Cambridge, 2001)
18. J. Williams, L. Hall, Aerodynamic sound generation by turbulent flow in the vicinity of a scattering half plane. J. Fluid Mech. **40**(4), 657–670 (1970)
19. N. Curle, The influence of solid boundaries upon aerodynamic sound. Proc. R. Soc. A **231**(1187), 505–514 (1955)
20. M. Potter, J. Foss, *Fluid Mechanics* (Great Lakes Press, Okemos, MI, 1982)
21. S. Hoerner, *Fluid-Dynamic Drag* (Self-Published, New York, 1965)
22. S. Yarusevych, P. Sullivan, J. Kawall, On vortex shedding from an airfoil in low-Reynolds-number flows. J. Fluid Mech. **632**, 245–271 (2009)

Generalizing the Butterfly Structure of the FFT

John Polcari

1 Introduction

The structure of the various forms of the fast Fourier transform (FFT) is typically described in terms of "butterfly" patterns [1], each involving only an individual pair of inputs or intermediate results. Here, the structure underlying the FFT is found to hold in a more general context applicable to any arbitrary radix-2 (i.e., $2^N \times 2^N$) complex unitary matrix. This is accomplished by identifying a decomposition of the matrix that leads to an equivalent generalized butterfly structure for which the FFT is then a special case. This procedure, known as multi-layer decomposition (MLD), effectively defines the underlying set of 2×2 complex Givens rotation matrices (each defined by a rotation angle and a separate phase angle) necessary to construct the original matrix. It is then possible to interpret each individual Givens rotation as a butterfly, permitting direct translation into the generalized butterfly structure shown in Fig. 1.

The critical mathematical insight leading to the generalized butterfly structure is that a radix-2 unitary matrix can always be factored using singular value decomposition (SVD) into a form where the central component takes the form of a layer of 2^{N-1} complex Givens rotations, while the left and right components are a total of four half size unitary matrices. In the process, the close conceptual relationship between MLD and SVD techniques is exposed. Recursive use of the same factorization on the components then ultimately results in Fig. 1.

J. Polcari (✉)
Informative Interpretations LLC, Burke, VA, USA

Oak Ridge National Laboratory, Oak Ridge, TN, USA
e-mail: polcarij@ornl.gov

A. A. Ruffa, B. Toni (eds.), *Advanced Research in Naval Engineering*, STEAM-H: Science, Technology, Engineering, Agriculture, Mathematics & Health, https://doi.org/10.1007/978-3-319-95117-1_3

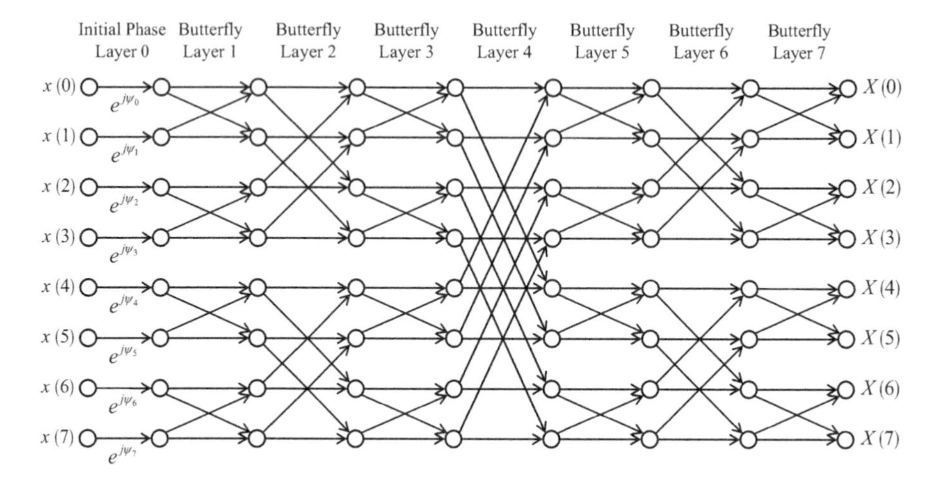

Fig. 1 Flow graph of fully decomposed eight element unitary transformation

Following some preliminary developments on standardized complex Givens rotations in Sect. 2, the central layer factorization is developed in Sect. 3. Section 4 then extends this procedure to provide the full MLD, while Sect. 5 provides details of the available MLD MATLAB code set. These sections are intended to provide an orderly introduction to MLD techniques; the reader is referred to [2] for a fully rigorous mathematical development of the approach.

While the flow graph in Fig. 1 is not inherently "fast" (effectively requiring the same operation count as required by a brute force matrix multiply), it is modestly more memory efficient, requiring storage of only one half the number of real values otherwise required. In addition, the composite matrix is always guaranteed to be unitary for any choice of rotation and phase parameters. Section 6 investigates the numerical symmetries that permit the generalized butterfly structure to collapse into the computationally efficient FFT.

Currently, MLD is more of a mathematical curiosity than a mainstream computational technique, and one that is not really even particularly widely recognized. A primary motivation for this chapter is to increase exposure to this elegant set of computational mathematics, in the belief that it may provide a key to a range of future advanced applications. In the summary found in Sect. 7, a number of different possible directions for future investigation and application are highlighted. Some of the proposed uses include:

- Design of "fast" approximate transformations, including speed/accuracy trade-offs.
- Re-orthogonalization of discretely sampled forms of continuous unitary functions.

- Synthetic unitary matrix multiplication and manipulation.
- Decimation-based approaches to SVD computation.

The analysis in this development is cast strictly in terms of radix-2 matrices, with the implication that other size matrices would be augmented by identity extension to fill them out to the required size. While the author conjectures that equivalent results for other choices of radix exist (as known to exist for the FFT), no effort is made here to explore that particular issue.

2 The Standardized Complex Givens Rotation

We begin by considering the required form of a 2×2 complex unitary matrix, as doing so exposes the standardized complex Givens rotation that is the essential building block encountered throughout this paper. Any arbitrary 2×2 unitary matrix may be written as

$$\underline{\overline{\mathbf{U}}} = \begin{bmatrix} |a|\, e^{j\psi_a} & |b|\, e^{j\varphi_b} \\ |c|\, e^{j\varphi_c} & |d|\, e^{j\psi_d} \end{bmatrix}. \tag{1}$$

However, it is useful to rewrite this in a phase normalized form by factoring out the phases of the main diagonal terms, so that

$$\underline{\overline{\mathbf{U}}} = \begin{bmatrix} |a|\, e^{j\psi_a} & |b|\, e^{j\varphi_b} \\ |c|\, e^{j\varphi_c} & |d|\, e^{j\psi_d} \end{bmatrix} = \begin{bmatrix} |a| & |b|\, e^{j(\varphi_b - \psi_d)} \\ |c|\, e^{j(\varphi_c - \psi_a)} & |d| \end{bmatrix} \begin{bmatrix} e^{j\psi_a} & 0 \\ 0 & e^{j\psi_d} \end{bmatrix}. \tag{2}$$

The properties of a unitary matrix [3] then produce the scalar equation sets

$$\underline{\overline{\mathbf{U}}}\,\underline{\overline{\mathbf{U}}}^{+} = \underline{\overline{\mathbf{I}}} \quad \Rightarrow \quad \begin{cases} |a|^2 + |b|^2 = 1 \\ |a|\,|c|\, e^{-j(\varphi_c - \psi_a)} + |b|\,|d|\, e^{j(\varphi_b - \psi_d)} = 0 \\ |a|\,|c|\, e^{j(\varphi_c - \psi_a)} + |b|\,|d|\, e^{-j(\varphi_b - \psi_d)} = 0 \\ |c|^2 + |d|^2 = 1 \end{cases} \tag{3}$$

and

$$\underline{\overline{\mathbf{U}}}^{+}\underline{\overline{\mathbf{U}}} = \underline{\overline{\mathbf{I}}} \quad \Rightarrow \quad \begin{cases} |a|^2 + |c|^2 = 1 \\ |a|\,|b|\, e^{j(\varphi_b - \psi_d)} + |c|\,|d|\, e^{-j(\varphi_c - \psi_a)} = 0 \\ |a|\,|b|\, e^{-j(\varphi_b - \psi_d)} + |c|\,|d|\, e^{j(\varphi_c - \psi_a)} = 0 \\ |b|^2 + |d|^2 = 1 \end{cases} \tag{4}$$

which, in turn, easily reduce to the amplitude requirement that

$$|a| = |d| = \cos\theta \quad \text{and} \quad |b| = |c| = \sin\theta \tag{5}$$

and the phase requirement that

$$e^{j(\varphi_b - \psi_d)} + e^{-j(\varphi_c - \psi_a)} = 0 \quad \Rightarrow \quad \varphi_c - \psi_a = \phi \;\; \text{and} \;\; \varphi_b - \psi_d = \pi - \phi. \tag{6}$$

As a result, any 2×2 phase normalized unitary matrix must take the form

$$\underline{\overline{\Gamma}}(\theta, \phi) = \begin{bmatrix} \cos\theta & -e^{-j\phi}\sin\theta \\ e^{j\phi}\sin\theta & \cos\theta \end{bmatrix}. \tag{7}$$

This matrix form may be interpreted as the complex extension of a 2×2 Givens rotation, the original real form of which is defined in [4]. Interpreted as a flow graph, it takes on the basic butterfly structure shown in Fig. 2.

Note that this differs in some details from the typical butterfly encountered in the flow graph associated with most FFTs; for example, the operation represented in Fig. 3 (reproduced from ([1], Fig. 6.6)), may be written as

$$\underline{\overline{\mathbf{B}}} = \begin{bmatrix} 1 & 1 \\ 1 & -1 \end{bmatrix} = \underline{\overline{\Gamma}}(\pi/4, 0)\left(\sqrt{2}\begin{bmatrix} 1 & 0 \\ 0 & -1 \end{bmatrix}\right). \tag{8}$$

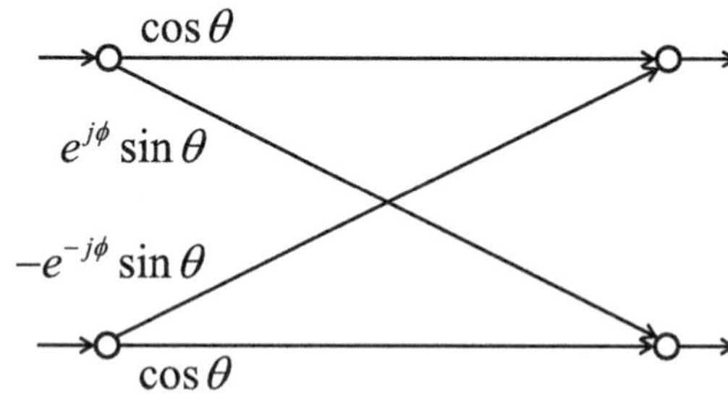

Fig. 2 Flow graph of individual complex Givens rotation butterfly

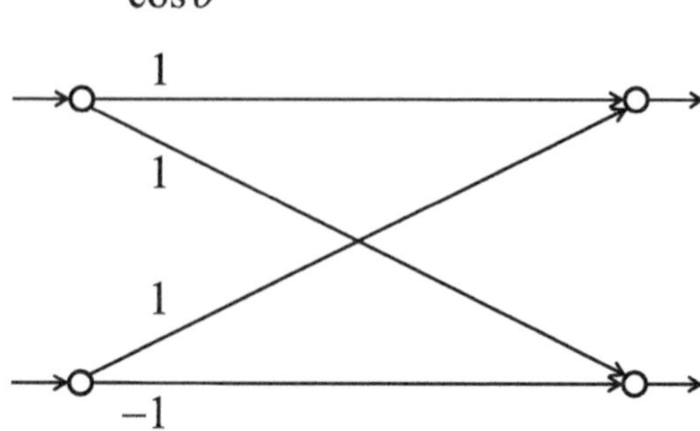

Fig. 3 Flow graph of a two-point fast Fourier transform

Hence, it may also be interpreted as a Givens rotation, but one cast in a form that is de-normalized in both amplitude and phase. While this de-normalization can certainly be useful in the absolute minimization of computational operations, it adds more confusion than insight when considering the underlying computational structure; hence, the standard butterfly structure is assumed to represent a standard complex Givens rotation, with the understanding that subsequent de-normalization is always possible should it lead to improved computational efficiency.

3 Unitary Matrix Central Layer Factorization

The critical mathematical insight leading to the generalized butterfly structure is that a $2M \times 2M$ unitary matrix can be factored into a form where the central component takes the form of a layer of M standard complex Givens rotations, while the left and right components are a total of four smaller $M \times M$ unitary matrices. In this section, a factorization of this type that applies to *any* arbitrary unitary matrix is derived. Interestingly, the procedure used is a matrix dual of that used to identify the essential Givens rotation structure in the previous section. In the process, the intimate mathematical relationship between butterfly structures and the SVD is also exposed.

Let $\underline{\overline{\mathbf{U}}}$ be a complex $2M \times 2M$ unitary matrix, so that it possesses the property

$$\underline{\overline{\mathbf{U}}}^{+}\underline{\overline{\mathbf{U}}} = \underline{\overline{\mathbf{U}}}\,\underline{\overline{\mathbf{U}}}^{+} = \underline{\overline{\mathbf{I}}}. \tag{9}$$

For reference, the generalized rotation represented by the unitary matrix operator $\underline{\mathbf{X}} = \underline{\overline{\mathbf{U}}}\ \underline{\mathbf{x}}$ may be expressed in terms of the flow graph representation shown in Fig. 4.

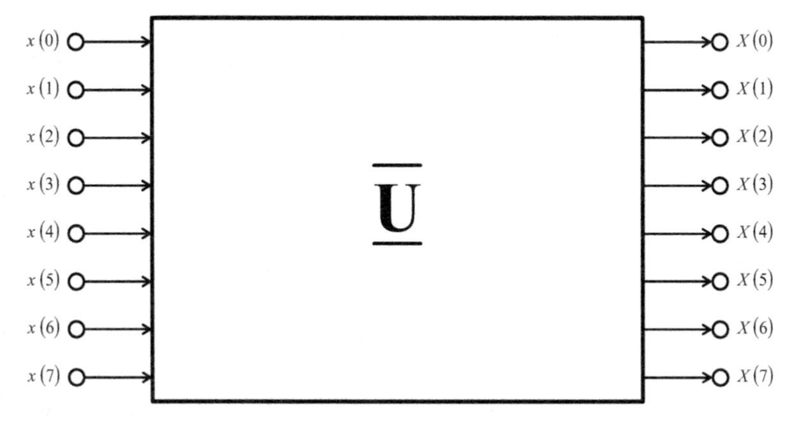

Fig. 4 Flow graph of arbitrary eight element unitary transformation

To begin the factorization, define the block matrix form of $\overline{\underline{\mathbf{U}}}$ as

$$\overline{\underline{\mathbf{U}}} = \begin{bmatrix} \overline{\underline{\mathbf{U}}}_{TL} & -\overline{\underline{\mathbf{U}}}_{TR} \\ \overline{\underline{\mathbf{U}}}_{BL} & \overline{\underline{\mathbf{U}}}_{BR} \end{bmatrix}, \tag{10}$$

where the minus sign in front of $\overline{\underline{\mathbf{U}}}_{TR}$ is included for notational convenience. None of the component blocks are unitary in and of themselves, but, since $\overline{\underline{\mathbf{U}}}$ is unitary, both

$$\begin{bmatrix} \overline{\underline{\mathbf{U}}}_{TL} & -\overline{\underline{\mathbf{U}}}_{TR} \\ \overline{\underline{\mathbf{U}}}_{BL} & \overline{\underline{\mathbf{U}}}_{BR} \end{bmatrix} \begin{bmatrix} \overline{\underline{\mathbf{U}}}^{+}_{TL} & \overline{\underline{\mathbf{U}}}^{+}_{BL} \\ -\overline{\underline{\mathbf{U}}}^{+}_{TR} & \overline{\underline{\mathbf{U}}}^{+}_{BR} \end{bmatrix} = \begin{bmatrix} \overline{\underline{\mathbf{I}}} & \overline{\underline{\mathbf{0}}} \\ \overline{\underline{\mathbf{0}}} & \overline{\underline{\mathbf{I}}} \end{bmatrix} \Rightarrow \begin{array}{l} \overline{\underline{\mathbf{U}}}_{TL}\overline{\underline{\mathbf{U}}}^{+}_{TL} + \overline{\underline{\mathbf{U}}}_{TR}\overline{\underline{\mathbf{U}}}^{+}_{TR} = \overline{\underline{\mathbf{I}}} \\ \overline{\underline{\mathbf{U}}}_{TL}\overline{\underline{\mathbf{U}}}^{+}_{BL} - \overline{\underline{\mathbf{U}}}_{TR}\overline{\underline{\mathbf{U}}}^{+}_{BR} = \overline{\underline{\mathbf{0}}} \\ \overline{\underline{\mathbf{U}}}_{BL}\overline{\underline{\mathbf{U}}}^{+}_{TL} - \overline{\underline{\mathbf{U}}}_{BR}\overline{\underline{\mathbf{U}}}^{+}_{TR} = \overline{\underline{\mathbf{0}}} \\ \overline{\underline{\mathbf{U}}}_{BL}\overline{\underline{\mathbf{U}}}^{+}_{BL} + \overline{\underline{\mathbf{U}}}_{BR}\overline{\underline{\mathbf{U}}}^{+}_{BR} = \overline{\underline{\mathbf{I}}} \end{array} \tag{11}$$

and

$$\begin{bmatrix} \overline{\underline{\mathbf{U}}}^{+}_{TL} & \overline{\underline{\mathbf{U}}}^{+}_{BL} \\ -\overline{\underline{\mathbf{U}}}^{+}_{TR} & \overline{\underline{\mathbf{U}}}^{+}_{BR} \end{bmatrix} \begin{bmatrix} \overline{\underline{\mathbf{U}}}_{TL} & -\overline{\underline{\mathbf{U}}}_{TR} \\ \overline{\underline{\mathbf{U}}}_{BL} & \overline{\underline{\mathbf{U}}}_{BR} \end{bmatrix} = \begin{bmatrix} \overline{\underline{\mathbf{I}}} & \overline{\underline{\mathbf{0}}} \\ \overline{\underline{\mathbf{0}}} & \overline{\underline{\mathbf{I}}} \end{bmatrix} \Rightarrow \begin{array}{l} \overline{\underline{\mathbf{U}}}^{+}_{TL}\overline{\underline{\mathbf{U}}}_{TL} + \overline{\underline{\mathbf{U}}}^{+}_{BL}\overline{\underline{\mathbf{U}}}_{BL} = \overline{\underline{\mathbf{I}}} \\ \overline{\underline{\mathbf{U}}}^{+}_{BL}\overline{\underline{\mathbf{U}}}_{BR} - \overline{\underline{\mathbf{U}}}^{+}_{TL}\overline{\underline{\mathbf{U}}}_{TR} = \overline{\underline{\mathbf{0}}} \\ \overline{\underline{\mathbf{U}}}^{+}_{BR}\overline{\underline{\mathbf{U}}}_{BL} - \overline{\underline{\mathbf{U}}}^{+}_{TR}\overline{\underline{\mathbf{U}}}_{TL} = \overline{\underline{\mathbf{0}}} \\ \overline{\underline{\mathbf{U}}}^{+}_{TR}\overline{\underline{\mathbf{U}}}_{TR} + \overline{\underline{\mathbf{U}}}^{+}_{BR}\overline{\underline{\mathbf{U}}}_{BR} = \overline{\underline{\mathbf{I}}} \end{array} \tag{12}$$

must hold. Now consider the SVD of each block, which is (presciently) labeled as

$$\begin{array}{ll} \overline{\underline{\mathbf{U}}}_{TL} = \overline{\underline{\mathbf{Y}}}_{T}\overline{\underline{\mathbf{\Lambda}}}_{C}\overline{\underline{\mathbf{Z}}}^{+}_{L} & \overline{\underline{\mathbf{U}}}_{TR} = \overline{\underline{\mathbf{Y}}}_{TR}\overline{\underline{\mathbf{\Lambda}}}_{TR}\overline{\underline{\mathbf{Z}}}^{+}_{TR} \\ \overline{\underline{\mathbf{U}}}_{BL} = \overline{\underline{\mathbf{Y}}}_{BL}\overline{\underline{\mathbf{\Lambda}}}_{BL}\overline{\underline{\mathbf{Z}}}^{+}_{BL} & \overline{\underline{\mathbf{U}}}_{BR} = \overline{\underline{\mathbf{Y}}}_{B}\overline{\underline{\mathbf{\Lambda}}}_{BR}\overline{\underline{\mathbf{Z}}}^{+}_{R}. \end{array} \tag{13}$$

Here the $\overline{\underline{\mathbf{Y}}}_{i}$ and $\overline{\underline{\mathbf{Z}}}_{i}$ are all $M \times M$ unitary singular vector matrices, while the $\overline{\underline{\mathbf{\Lambda}}}_{i}$ are $M \times M$ real non-negative diagonal singular value matrices. Replacing the component blocks in the first equation of (11) by their respective SVD representations yields

$$\overline{\underline{\mathbf{Y}}}_{T}\overline{\underline{\mathbf{\Lambda}}}^{2}_{C}\overline{\underline{\mathbf{Y}}}^{+}_{T} + \overline{\underline{\mathbf{Y}}}_{TR}\overline{\underline{\mathbf{\Lambda}}}^{2}_{TR}\overline{\underline{\mathbf{Y}}}^{+}_{TR} = \overline{\underline{\mathbf{I}}} \Rightarrow \overline{\underline{\mathbf{\Lambda}}}^{2}_{C} + \overline{\underline{\mathbf{Y}}}^{+}_{T}\overline{\underline{\mathbf{Y}}}_{TR}\overline{\underline{\mathbf{\Lambda}}}^{2}_{TR}\overline{\underline{\mathbf{Y}}}^{+}_{TR}\overline{\underline{\mathbf{Y}}}_{T} = \overline{\underline{\mathbf{I}}}. \tag{14}$$

As both $\overline{\underline{\mathbf{\Lambda}}}^{2}_{C}$ and $\overline{\underline{\mathbf{I}}}$ are diagonal, the second result in (14) requires that

$$\overline{\underline{\mathbf{Y}}}^{+}_{T}\overline{\underline{\mathbf{Y}}}_{TR} = \overline{\underline{\mathbf{\Phi}}}_{T} \Rightarrow \overline{\underline{\mathbf{Y}}}_{TR} = \overline{\underline{\mathbf{Y}}}_{T}\overline{\underline{\mathbf{\Phi}}}_{T}, \tag{15}$$

where $\overline{\underline{\mathbf{\Phi}}}_{T}$ is necessarily diagonal. Further, since both $\overline{\underline{\mathbf{Y}}}_{T}$ and $\overline{\underline{\mathbf{Y}}}_{TR}$ are unitary, $\overline{\underline{\mathbf{\Phi}}}_{T}$ must also be unitary, it must then take on the form of an $M \times M$ diagonal matrix with unit magnitude complex phasors arrayed down the diagonal, that is

$$\underline{\overline{\boldsymbol{\Phi}}}_T = \begin{bmatrix} e^{j\phi_{T0}} & \cdots & 0 \\ \vdots & \ddots & \vdots \\ 0 & \cdots & e^{j\phi_{T(M-1)}} \end{bmatrix}. \tag{16}$$

As a consequence, (14) then reduces to

$$\underline{\overline{\boldsymbol{\Lambda}}}_C^2 + \underline{\overline{\boldsymbol{\Lambda}}}_{TR}^2 = \underline{\overline{\mathbf{I}}}. \tag{17}$$

Repeating this analysis on the last equation in (11) and the first and last equations in (12) leads to the requirements that

$$\begin{array}{ll} \underline{\overline{\mathbf{Y}}}_{TR} = \underline{\overline{\mathbf{Y}}}_T \underline{\overline{\boldsymbol{\Phi}}}_T & \underline{\overline{\mathbf{Z}}}_{BL} = \underline{\overline{\mathbf{Z}}}_L \underline{\overline{\boldsymbol{\Phi}}}_L \\ \underline{\overline{\mathbf{Y}}}_{BL} = \underline{\overline{\mathbf{Y}}}_B \underline{\overline{\boldsymbol{\Phi}}}_B & \underline{\overline{\mathbf{Z}}}_{TR} = \underline{\overline{\mathbf{Z}}}_R \underline{\overline{\boldsymbol{\Phi}}}_R \end{array} \tag{18}$$

and

$$\begin{array}{ll} \underline{\overline{\boldsymbol{\Lambda}}}_C^2 + \underline{\overline{\boldsymbol{\Lambda}}}_{TR}^2 = \underline{\overline{\mathbf{I}}} & \underline{\overline{\boldsymbol{\Lambda}}}_C^2 + \underline{\overline{\boldsymbol{\Lambda}}}_{BL}^2 = \underline{\overline{\mathbf{I}}} \\ \underline{\overline{\boldsymbol{\Lambda}}}_{BL}^2 + \underline{\overline{\boldsymbol{\Lambda}}}_{BR}^2 = \underline{\overline{\mathbf{I}}} & \underline{\overline{\boldsymbol{\Lambda}}}_{TR}^2 + \underline{\overline{\boldsymbol{\Lambda}}}_{BR}^2 = \underline{\overline{\mathbf{I}}} \end{array}, \tag{19}$$

which, since all the various singular values are non-negative, further implies that

$$\underline{\overline{\boldsymbol{\Lambda}}}_{BR} = \underline{\overline{\boldsymbol{\Lambda}}}_C \quad \text{and} \quad \underline{\overline{\boldsymbol{\Lambda}}}_{TR} = \underline{\overline{\boldsymbol{\Lambda}}}_{BL} = \underline{\overline{\boldsymbol{\Lambda}}}_S \quad \text{with} \quad \underline{\overline{\boldsymbol{\Lambda}}}_C^2 + \underline{\overline{\boldsymbol{\Lambda}}}_S^2 = \underline{\overline{\mathbf{I}}}. \tag{20}$$

Note that the final result in (20) requires all the singular values in both $\underline{\overline{\boldsymbol{\Lambda}}}_C$ and $\underline{\overline{\boldsymbol{\Lambda}}}_S$ to not exceed unity, and, assuming that the singular values in $\underline{\overline{\boldsymbol{\Lambda}}}_C$ are arranged in the conventional order of decreasing magnitude, those in $\underline{\overline{\boldsymbol{\Lambda}}}_S$ must be arranged in the opposite order of increasing magnitude.

Define the additional diagonal phase matrices

$$\underline{\overline{\boldsymbol{\Phi}}} = \underline{\overline{\boldsymbol{\Phi}}}_B \underline{\overline{\boldsymbol{\Phi}}}_L^+ \quad \text{and} \quad \underline{\overline{\boldsymbol{\Phi}}}_{TR} = \underline{\overline{\boldsymbol{\Phi}}}_T \underline{\overline{\boldsymbol{\Phi}}}_R^+; \tag{21}$$

as the various diagonal matrices all commute, the set of block SVDs then reduce to

$$\begin{array}{ll} \underline{\overline{\mathbf{U}}}_{TL} = \underline{\overline{\mathbf{Y}}}_T \underline{\overline{\boldsymbol{\Lambda}}}_C \underline{\overline{\mathbf{Z}}}_L^+ & \underline{\overline{\mathbf{U}}}_{TR} = \underline{\overline{\mathbf{Y}}}_T \underline{\overline{\boldsymbol{\Lambda}}}_S \underline{\overline{\boldsymbol{\Phi}}}_{TR} \underline{\overline{\mathbf{Z}}}_R^+ \\ \underline{\overline{\mathbf{U}}}_{BL} = \underline{\overline{\mathbf{Y}}}_B \underline{\overline{\boldsymbol{\Lambda}}}_S \underline{\overline{\boldsymbol{\Phi}}}\, \underline{\overline{\mathbf{Z}}}_L^+ & \underline{\overline{\mathbf{U}}}_{BR} = \underline{\overline{\mathbf{Y}}}_B \underline{\overline{\boldsymbol{\Lambda}}}_C \underline{\overline{\mathbf{Z}}}_R^+. \end{array} \tag{22}$$

Substituting these results into the second equation in (11) yields[1]

$$\underline{\overline{\mathbf{Y}}}_T \underline{\overline{\boldsymbol{\Lambda}}}_C \underline{\overline{\boldsymbol{\Phi}}}^+ \underline{\overline{\boldsymbol{\Lambda}}}_S \underline{\overline{\mathbf{Y}}}_B^+ - \underline{\overline{\mathbf{Y}}}_T \underline{\overline{\boldsymbol{\Lambda}}}_S \underline{\overline{\boldsymbol{\Phi}}}_{TR} \underline{\overline{\boldsymbol{\Lambda}}}_C \underline{\overline{\mathbf{Y}}}_B^+ = \underline{\overline{\mathbf{0}}} \quad \Rightarrow \quad \underline{\overline{\boldsymbol{\Phi}}}_{TR} = \underline{\overline{\boldsymbol{\Phi}}}^+, \tag{23}$$

[1]The other three middle equations in (11) and (12) lead to results that duplicate those of (23).

Then the original unitary matrix may be written in factored form as

$$\underline{\overline{\mathbf{U}}}=\begin{bmatrix} \underline{\overline{\mathbf{Y}}}_T\underline{\overline{\boldsymbol{\Lambda}}}_C\underline{\overline{\mathbf{Z}}}_L^+ & -\underline{\overline{\mathbf{Y}}}_T\underline{\overline{\boldsymbol{\Lambda}}}_S\underline{\overline{\boldsymbol{\Phi}}}^+\underline{\overline{\mathbf{Z}}}_R^+ \\ \underline{\overline{\mathbf{Y}}}_B\underline{\overline{\boldsymbol{\Lambda}}}_S\underline{\overline{\boldsymbol{\Phi}}}\,\underline{\overline{\mathbf{Z}}}_L^+ & \underline{\overline{\mathbf{Y}}}_B\underline{\overline{\boldsymbol{\Lambda}}}_C\underline{\overline{\mathbf{Z}}}_R^+ \end{bmatrix}=\begin{bmatrix} \underline{\overline{\mathbf{Y}}}_T & \underline{\overline{\mathbf{0}}} \\ \underline{\overline{\mathbf{0}}} & \underline{\overline{\mathbf{Y}}}_B \end{bmatrix}\begin{bmatrix} \underline{\overline{\boldsymbol{\Lambda}}}_C & -\underline{\overline{\boldsymbol{\Lambda}}}_S\underline{\overline{\boldsymbol{\Phi}}}^+ \\ \underline{\overline{\boldsymbol{\Lambda}}}_S\underline{\overline{\boldsymbol{\Phi}}} & \underline{\overline{\boldsymbol{\Lambda}}}_C \end{bmatrix}\begin{bmatrix} \underline{\overline{\mathbf{Z}}}_L & \underline{\overline{\mathbf{0}}} \\ \underline{\overline{\mathbf{0}}} & \underline{\overline{\mathbf{Z}}}_R \end{bmatrix}^+, \tag{24}$$

and the desired factorization has been achieved.

Now consider the required form of the central component matrix. As mentioned previously, none of the singular values in either $\underline{\overline{\boldsymbol{\Lambda}}}_C$ or $\underline{\overline{\boldsymbol{\Lambda}}}_S$ can exceed unity; it is then convenient to parameterize $\underline{\overline{\boldsymbol{\Lambda}}}_C$ in terms of the angle set

$$\underline{\overline{\boldsymbol{\Lambda}}}_C=\begin{bmatrix} \cos\theta_0 & \cdots & 0 \\ \vdots & \ddots & \vdots \\ 0 & \cdots & \cos\theta_{M-1} \end{bmatrix} \Rightarrow \underline{\overline{\boldsymbol{\Lambda}}}_S=\begin{bmatrix} \sin\theta_0 & \cdots & 0 \\ \vdots & \ddots & \vdots \\ 0 & \cdots & \sin\theta_{M-1} \end{bmatrix}. \tag{25}$$

As a result

$$\begin{bmatrix} \underline{\overline{\boldsymbol{\Lambda}}}_C & -\underline{\overline{\boldsymbol{\Lambda}}}_S\underline{\overline{\boldsymbol{\Phi}}}^+ \\ \underline{\overline{\boldsymbol{\Lambda}}}_S\underline{\overline{\boldsymbol{\Phi}}} & \underline{\overline{\boldsymbol{\Lambda}}}_C \end{bmatrix}$$

$$=\begin{bmatrix} \cos\theta_0 & \cdots & 0 & -e^{-j\phi_0}\sin\theta_0 & \cdots & 0 \\ \vdots & \ddots & \vdots & \vdots & \ddots & \vdots \\ 0 & \cdots & \cos\theta_{M-1} & 0 & \cdots & -e^{-j\phi_{M-1}}\sin\theta_{M-1} \\ e^{j\phi_0}\sin\theta_0 & \cdots & 0 & \cos\theta_0 & \cdots & 0 \\ \vdots & \ddots & \vdots & \vdots & \ddots & \vdots \\ 0 & \cdots & e^{j\phi_{M-1}}\sin\theta_{M-1} & 0 & \cdots & \cos\theta_{M-1} \end{bmatrix}, \tag{26}$$

and extracting the row/column pair $(i, M+i)$ yields the standard form of a 2×2 complex Givens rotation matrix

$$\underline{\overline{\boldsymbol{\Gamma}}}(\theta_i, \phi_i)=\begin{bmatrix} \cos\theta_i & -e^{-j\phi_i}\sin\theta_i \\ e^{j\phi_i}\sin\theta_i & \cos\theta_i \end{bmatrix}. \tag{27}$$

Expressing these results in the vernacular of a flow graph, the computational operator has been converted from the full unitary matrix shown in Fig. 4 to the hybrid structure shown in Fig. 5, where the center layer is now constructed of M parallel individual butterfly operations from Fig. 2.

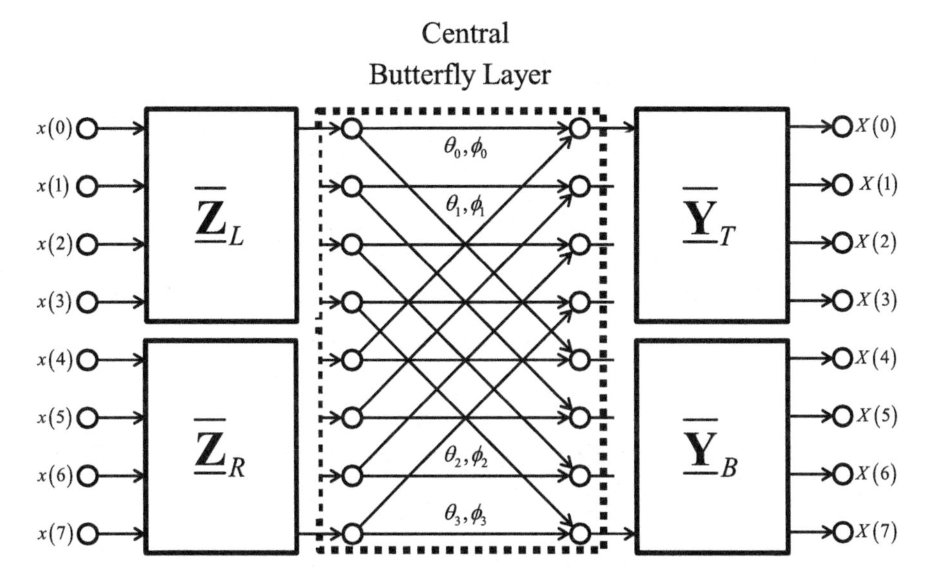

Fig. 5 Flow graph of single layer decomposed eight element unitary transformation

4 Full Unitary Matrix Decomposition

Extending these results to provide a complete butterfly representation of any radix-2 unitary matrix is quite simple; the above factorization is now recursively employed to expose the central layers of each of the four unitary matrices comprising the left and right components, as shown in Fig. 6.

This procedure may be continued until the unitary matrices are reduced to 2×2 size. Then, as previously shown in (2), each of these 2×2 matrices may be represented as one more Givens rotation multiplied by a diagonal matrix consisting of two residual complex phase scalars along the diagonal. It turns out that all the phase residuals can be collected into a single set of 2^N phase adjustments that may be implemented at any point in the flow graph. While formal proof of this result is mathematically tedious (involving much cumbersome notation to precisely track the associated phase accounting), the basic concept is quite simple. The realization that

$$\begin{bmatrix} e^{j\psi_0} & 0 \\ 0 & e^{j\psi_1} \end{bmatrix} \underline{\overline{\Gamma}}(\theta, \phi) = \underline{\overline{\Gamma}}(\theta, \phi + \psi_1 - \psi_0) \begin{bmatrix} e^{j\psi_0} & 0 \\ 0 & e^{j\psi_1} \end{bmatrix} \tag{28}$$

provides a method for propagating the phase residuals across the various flow graph layers to the point of collection, where they may be accumulated into composite phase adjustments. This leads to the generalized butterfly structure shown in Fig. 7, where, by convention, the composite adjustments are located at the beginning of the graph.

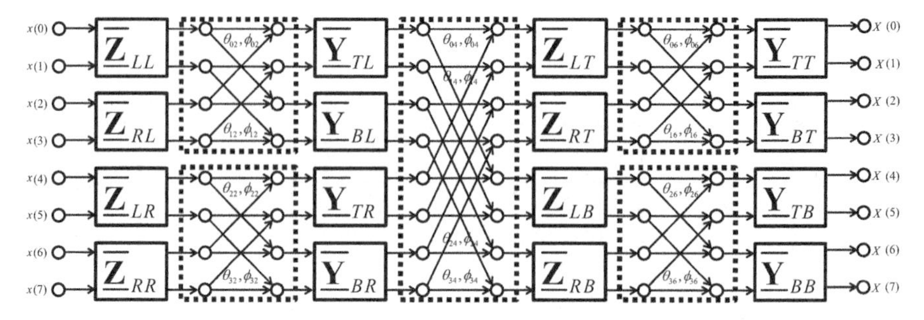

Fig. 6 Flow graph of partially expanded eight element unitary transformation

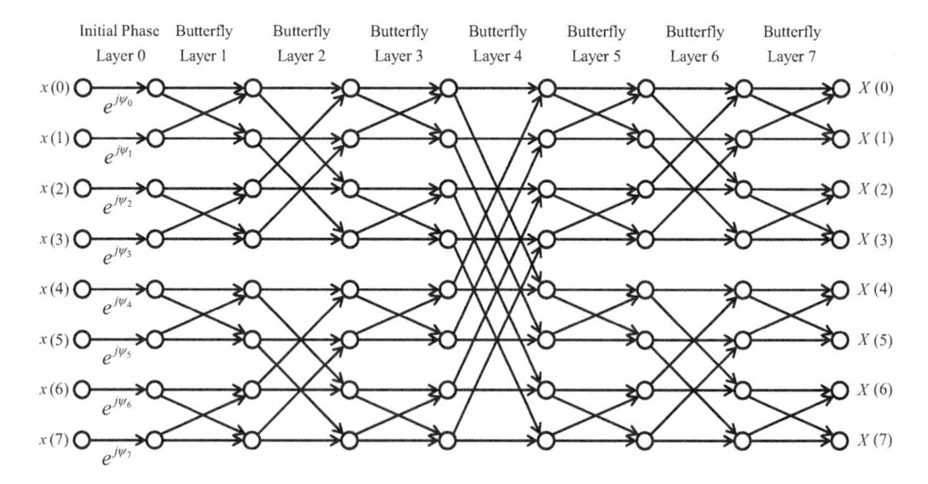

Fig. 7 Flow graph of fully decomposed eight element unitary transformation

The fully generalized butterfly structure is comprised of this initial phase layer followed by $2^N - 1$ butterfly layers, each made up of 2^{N-1} Givens butterflies operating in parallel. The stride of each layer (that is, the distance between the elements connected by each individual butterfly) varies from layer to layer, in keeping with the pattern

$$
\begin{gathered}
1, 2, 1, 4, 1, 2, 1, 8, 1, 2, 1, 4, 1, \cdots \\
1, 2, 1, 2^{N-1}, 1, 2, 1, \cdots \\
1, 2, 1, 4, 1, 2, 1, 8, 1, 2, 1, 4, 1,
\end{gathered} \tag{29}
$$

A useful geometric interpretation of this result is that it defines the precise sequence of individual planar coordinate rotations needed to implement the general 2^N dimensional coordinate rotation defined by the unitary matrix. It may also be considered the equivalent polar representation of a unitary matrix.

Note that the butterfly structure for the conjugate transpose matrix may be found simply by reversing the layer order and then negating each Givens rotation angle θ. This places phase adjustment as the final rather than initial layer, but, if desired, it can then be propagated back to the beginning by repeated use of (28).

A few comparisons of computational and storage efficiency are insightful. As might be anticipated, the fully generalized butterfly structure is not inherently "fast." Each of the $(2^N - 1)2^{N-1}$ butterflies requires two complex multiplies (for inner branch scaling), two real-complex multiplies (for outer branch scaling), and two complex adds (for output sums), or, equivalently, 12 real multiplies and 6 real adds. In addition, each of the 2^N initial phase adjustments requires 1 complex multiply, or, equivalently, 4 real multiplies and 2 real adds. Hence, for the standard normalized butterfly structure

$$\begin{aligned}\text{Real Multiplies} &= 12\left(2^{N-1}\right)\left(2^N - 1\right) + 4\left(2^N\right) = 2^{2N+2} + 2^{2N+1} - 2^{N+1} \\ \text{Real Adds} &= 6\left(2^{N-1}\right)\left(2^N - 1\right) + 2\left(2^N\right) = 2^{2N+1} + 2^{2N} - 2^N\end{aligned} \tag{30}$$

By comparison, brute force matrix-vector multiplication by the original complex $\underline{\overline{U}}$ matrix requires only 2^{2N+2} real multiplies but $2^{2N+2} - 2^{N+1}$ real adds. By amplitude de-normalization of the butterflies, the need for any outer branch scaling can be eliminated, with the scaling effects then being combined with the phase adjustments at no additional computational cost. This reduces real multiplication count to precisely match that of the brute force approach, while retaining the modest advantage in real addition count.

While the fully general butterfly structure provides at best marginal improvements in computational efficiency, it does provide some efficiencies in memory storage. The Givens rotation angles and phase angles together with the phase adjustment angles provide a total of only 2^{2N} real values, compared to original matrix, which requires storage of the same number of complex values (or twice as many real values).

5 Practical Generalized Butterfly Decomposition Codes

A set of MATLAB routines implementing the MLD algorithm can be found in [5]. This code provides three functional capabilities:

- It decomposes an arbitrary unitary matrix into its component butterflies, returning the 2^N phase adjustments in the initial layer and the rotation/phase angle pair for the 2^{N-1} individual butterflies in each of the $2^N - 1$ layers of the flow graph;
- It recreates the matrix from the flow graph specification; and
- It applies the flow graph to any specified set of length 2^N input column vectors.

The code set is written to provide accuracy and robustness at the potential expense of some computational speed. The critical practical issue involves the details of creating a normalized SVD result. While standard SVD routines are well designed in terms of their singular value outputs, their singular vector outputs are typically not carefully normalized. The following example, drawn from the standard MATLAB SVD routine, illustrates the point. For the arbitrarily chosen matrix

$$\overline{\underline{\mathbf{A}}} = \begin{bmatrix} 3\text{-}2j & 5+3j \\ 1+5j & 4+j \end{bmatrix},$$

the SVD routine returns an SVD of

$$\overline{\underline{\mathbf{U}}}_X \overline{\underline{\mathbf{\Lambda}}}_X \overline{\underline{\mathbf{V}}}_X^+ = \begin{bmatrix} \text{-}0.726\text{-}0.158j & \text{-}0.096\text{-}0.662j \\ \text{-}0.479\text{-}0.467j & \text{-}0.299+0.681j \end{bmatrix} \begin{bmatrix} 7.998 & 0 \\ 0 & 5.102 \end{bmatrix} \begin{bmatrix} \text{-}0.585 & 0.811 \\ \text{-}0.811 & \text{-}0.585 \end{bmatrix}^+ .$$

Now let $\overline{\underline{\mathbf{T}}}$ be the unitary matrix

$$\overline{\underline{\mathbf{T}}} = \frac{1}{\sqrt{2}} \begin{bmatrix} 1 & 1 \\ 1 & -1 \end{bmatrix}; \tag{31}$$

if the SVD routine were carefully normalized, then the SVD of the product $\overline{\underline{\mathbf{Y}}} = \overline{\underline{\mathbf{T}}}\,\overline{\underline{\mathbf{X}}}$ would be

$$\overline{\underline{\mathbf{U}}}_Y \overline{\underline{\mathbf{\Lambda}}}_Y \overline{\underline{\mathbf{V}}}_Y^+ = \left(\overline{\underline{\mathbf{T}}}\,\overline{\underline{\mathbf{U}}}_X\right) \overline{\underline{\mathbf{\Lambda}}}_X \overline{\underline{\mathbf{V}}}_X^+ \quad \Rightarrow \quad \begin{matrix} \overline{\underline{\mathbf{U}}}_Y = \overline{\underline{\mathbf{T}}}\,\overline{\underline{\mathbf{U}}}_X \\ \overline{\underline{\mathbf{\Lambda}}}_Y = \overline{\underline{\mathbf{\Lambda}}}_X \\ \overline{\underline{\mathbf{V}}}_Y = \overline{\underline{\mathbf{V}}}_X \end{matrix} . \tag{32}$$

However, in reality

$$\overline{\underline{\mathbf{U}}}_Y = \begin{bmatrix} 0.852+0.442j & -0.280+0.013j \\ 0.175-0.218j & 0.144-0.949j \end{bmatrix} \quad \text{with}$$

$$\overline{\underline{\mathbf{T}}}\,\overline{\underline{\mathbf{U}}}_X = \begin{bmatrix} -0.852-0.442j & -0.280+0.013j \\ -0.175+0.218j & 0.144-0.949j \end{bmatrix},$$

while

$$\overline{\underline{\mathbf{V}}}_Y = \begin{bmatrix} -0.585 & 0.811 \\ -0.811 & -0.585 \end{bmatrix} \quad \text{with} \quad \overline{\underline{\mathbf{V}}}_X = \begin{bmatrix} 0.585 & 0.811 \\ 0.811 & -0.585 \end{bmatrix}.$$

If one is focused primarily upon the singular values, the differences in the above results are essentially trivial; however, consistent singular vector normalization turns out to be a centerpiece of designing a practical MLD algorithm. From a code perspective, the implementation of such normalization turns out to be much more tedious than might be anticipated, adding significant notational load while shedding little useful insight. Proper handling of normalization for zero or near-zero singular values is particularly tedious. In the interest of brevity, this subject is not discussed here in further detail; a full discussion of SVD ambiguities may be found in [6, Sect. 4] and the specifics of the normalization process in [2, Sects. 4, 5, and Appendix C].

Three principal routines are provided in the multi-layer decomposition (MLD) package. Any radix-2 unitary matrix $\underline{\overline{\mathrm{U}}}$ may be decomposed by calling the routine

$$\mathrm{MLD} = \mathrm{mld}\,(\mathrm{U})\,.$$

Here, MLD is a MATLAB structure with three fields:

- R_N_L is the array of Givens rotation angles expressed in radians, one for each of the butterflies in each layer.
- P_N_L is the array of complex Givens phase angles expressed in radians, again one for each of the butterflies in each layer.
- Q_2N is the vector of initial complex phase adjustments expressed in radians, one for each input element.

The matrix may by reconstructed from the MLD structure by calling the routine

$$\mathrm{U} = \mathrm{assemblemld}\,(\mathrm{MLD})\,.$$

Finally, the matrix may be applied to a set of length 2^N input vectors using the flow graph formulation by calling the routine

$$\mathrm{Xo2NM} = \mathrm{applymld}\,(\mathrm{Xi2NM}, \mathrm{setupxmld}\,(\mathrm{MLD}))\,.$$

Here, Xi_2N_M and Xo_2N_M are, respectively, the input and output vector sets (the second dimension providing the option of multiple sets of vectors to be evaluated in parallel), with the setup_xmld() utility expanding the MLD structure to a form more convenient for the computation of the butterfly flow graph. The package also contains several additional subsidiary routines, for a total of ten code files.

As typical for methods that preserve orthogonality, the error characteristics of the MLD package are quite good. Table 1 provides the accuracies achieved in the matrix reconstruction resulting from these routines for various sizes of randomly selected unitary matrix.

Table 1 Accuracies achieved in multi-layer decomposition reconstruction

Matrix size	Average relative accuracy
4×4	6.1×10^{-16}
16×16	3.5×10^{-15}
64×64	1.0×10^{-14}
256×256	2.1×10^{-14}
1024×1024	4.5×10^{-14}

6 Recovering the FFT

We now consider how the traditional FFT emerges from this more generalized form of butterfly flow graph. In doing so, some important insights emerge regarding symmetries that lead to more general forms of computationally fast transformation. The FFT of length 2^N is traditionally taken to be a particular computational implementation of the analysis (time to frequency) discrete Fourier transform [1, Eq. (3.4)]

$$X(k) = \sum_{n=0}^{2^N-1} x(n) e^{-j(2\pi nk/2^N)}, \tag{33}$$

where the two length 2^N vectors $\underline{\mathbf{x}} = [x(n)]$ and $\underline{\mathbf{X}} = [X(k)]$ represent data series in the time and frequency domains, respectively. The time domain results are assumed to be arranged in the traditional bit-reversed order, so that, for an eight element FFT (which is used here for illustration purposes), the time vector is defined as

$$\underline{\mathbf{x}}_{br} = [x(0)\ \ x(4)\ \ x(2)\ \ x(6)\ \ x(1)\ \ x(5)\ \ x(3)\ \ x(7)]^T. \tag{34}$$

To obtain a unitary matrix formulation, it is convenient to adopt a symmetric form of normalization, so that both the FFT (analysis transform) and the inverse FFT (synthesis transform) carry scaling factors of $1/2^{N/2}$. Then the eight element analysis transform may be written in matrix-vector form as

$$\underline{\mathbf{X}} = \overline{\underline{\mathbf{F}}}_A \underline{\mathbf{x}}_{br}, \tag{35}$$

where

$$\overline{\underline{\mathbf{F}}}_A = \frac{1}{\sqrt{8}} \begin{bmatrix} f_{A00} & \cdots & f_{A07} \\ \vdots & \ddots & \vdots \\ f_{A70} & \cdots & f_{A77} \end{bmatrix} \quad \text{with} \quad f_{Akn} = e^{-j2\pi(k\,\mathrm{br}(n)/8)}. \tag{36}$$

This matrix is necessarily unitary, as the equivalent synthesis transform is

$$\underline{\mathbf{x}} = \overline{\underline{\mathbf{F}}}_S \underline{\mathbf{X}}_{br} = \overline{\underline{\mathbf{F}}}_A^{+} \underline{\mathbf{X}}_{br}, \tag{37}$$

and recovery of the original temporal data then requires

$$\overline{\underline{\mathbf{F}}}_S \overline{\underline{\mathbf{F}}}_A = \overline{\underline{\mathbf{F}}}_A^{+} \overline{\underline{\mathbf{F}}}_A = \underline{\mathbf{I}}. \tag{38}$$

The use of MLD on this matrix leads to the Givens butterfly structure and initial phase adjustments shown in Table 2. As is apparent, the Givens rotations associated with all the butterflies in layers 3 and 5–7 reduce to identity matrices, so that those

Table 2 Multi-layer decomposition butterfly parameters and initial phase adjustments for eight element decimation-in-time FFT

Butterfly rotation angle θ (radians)							
Layer	1	2	3	4	5	6	7
Butterfly 1	$\pi/4$	$\pi/4$	0	$\pi/4$	0	0	0
Butterfly 2	$\pi/4$	$\pi/4$	0	$\pi/4$	0	0	0
Butterfly 3	$\pi/4$	$\pi/4$	0	$\pi/4$	0	0	0
Butterfly 4	$\pi/4$	$\pi/4$	0	$\pi/4$	0	0	0
Butterfly phase angle ϕ (radians)							
Layer	1	2	3	4	5	6	7
Butterfly 1	0	0	0	0	0	0	0
Butterfly 2	$-\pi/2$	0	0	0	0	0	0
Butterfly 3	$-\pi/4$	$-\pi/2$	0	0	0	0	0
Butterfly 4	$-3\pi/4$	$-\pi/2$	0	0	0	0	0

Initial phase adjustment ψ (radians)								
Element	0	4	2	6	1	5	3	7
ψ	0	π	π	$-\pi/2$	π	$-\pi/4$	$-\pi/2$	$-\pi/4$

layers effectively become simple pass-through operations. As a result, the flow graph collapses to the traditional FFT butterfly pattern illustrated in Fig. 8. Note that the $1/\sqrt{8} = \left(1/\sqrt{2}\right)^3$ scaling factor represents precisely the de-normalization factor arising from each of the three layers of remaining butterflies.

Two observations regarding computational complexity reduction are germane. First, the reduction of the Givens rotation angle to any multiple of $\pi/2$ converts the butterfly to a pass-through or element-reversal operation. Similarly, reduction of the Givens rotation angle to any odd multiple of $\pi/4$ converts the butterfly to a sum-difference operation, with larger or smaller computational load reduction depending upon the specific value of the associated phase angle.

A less obvious approach for reducing computational complexity involves identification of the particular symmetry characteristics of the FFT that permit layer elimination. If the central layer in the unitary matrix consists of a set of matched Given rotations (that is, all possessing identical rotation angles), (24) takes on the form

$$
\begin{aligned}
\underline{\overline{\mathbf{U}}} &= \begin{bmatrix} \alpha \underline{\overline{\mathbf{Y}}}_T \underline{\overline{\mathbf{Z}}}_L^+ & -\sqrt{1-\alpha^2}\underline{\overline{\mathbf{Y}}}_T \underline{\overline{\mathbf{\Phi}}}^+ \underline{\overline{\mathbf{Z}}}_R^+ \\ \sqrt{1-\alpha^2}\underline{\overline{\mathbf{Y}}}_B \underline{\overline{\mathbf{\Phi}}} \underline{\overline{\mathbf{Z}}}_L^+ & \alpha \underline{\overline{\mathbf{Y}}}_B \underline{\overline{\mathbf{Z}}}_R^+ \end{bmatrix} \\
&= \begin{bmatrix} \underline{\overline{\mathbf{Y}}}_T \underline{\overline{\mathbf{Z}}}_L^+ & \underline{\overline{\mathbf{0}}} \\ \underline{\overline{\mathbf{0}}} & \underline{\overline{\mathbf{Y}}}_B \underline{\overline{\mathbf{Z}}}_R^+ \end{bmatrix} \begin{bmatrix} \alpha \underline{\overline{\mathbf{I}}} & -\sqrt{1-\alpha^2}\underline{\overline{\mathbf{Z}}}_L \underline{\overline{\mathbf{\Phi}}}^+ \underline{\overline{\mathbf{Z}}}_R^+ \\ \sqrt{1-\alpha^2}\underline{\overline{\mathbf{Z}}}_R \underline{\overline{\mathbf{\Phi}}}\, \underline{\overline{\mathbf{Z}}}_L^+ & \alpha \underline{\overline{\mathbf{I}}} \end{bmatrix} \\
&= \left[\begin{bmatrix} \alpha \underline{\overline{\mathbf{I}}} & -\sqrt{1-\alpha^2}\underline{\overline{\mathbf{Y}}}_T \underline{\overline{\mathbf{\Phi}}}^+ \underline{\overline{\mathbf{Y}}}_B^+ \\ \sqrt{1-\alpha^2}\underline{\overline{\mathbf{Y}}}_B \underline{\overline{\mathbf{\Phi}}}\, \underline{\overline{\mathbf{Y}}}_T^+ & \alpha \underline{\overline{\mathbf{I}}} \end{bmatrix} \begin{bmatrix} \underline{\overline{\mathbf{Z}}}_L \underline{\overline{\mathbf{Y}}}_T^+ & \underline{\overline{\mathbf{0}}} \\ \underline{\overline{\mathbf{0}}} & \underline{\overline{\mathbf{Z}}}_R \underline{\overline{\mathbf{Y}}}_B^+ \end{bmatrix} \right]^+ .
\end{aligned}
\tag{39}
$$

Then, if either

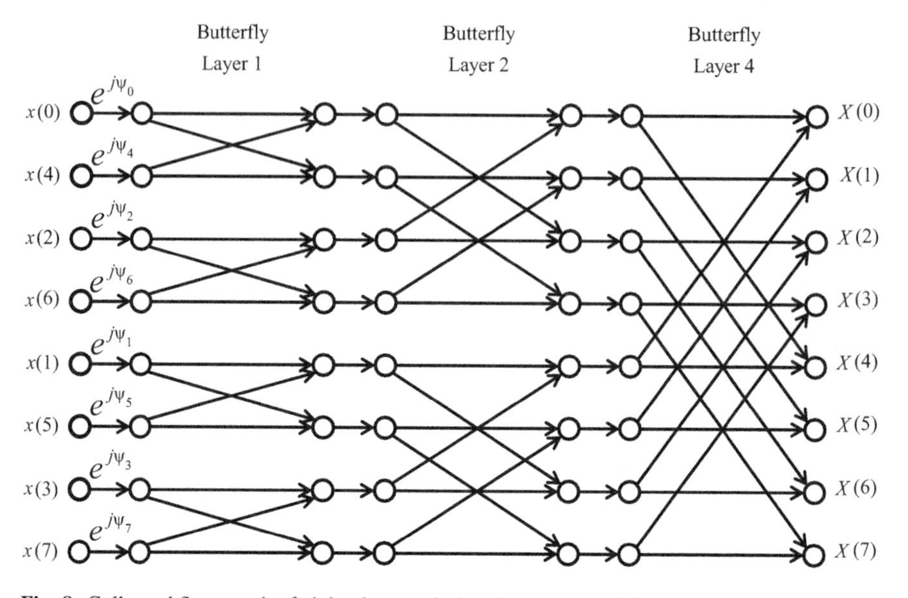

Fig. 8 Collapsed flow graph of eight element decimation-in-time FFT

$$
\begin{aligned}
&\overline{\underline{\mathbf{Z}}}_R \overline{\underline{\mathbf{\Phi}}}\, \overline{\underline{\mathbf{Z}}}_L^+ = \overline{\underline{\mathbf{\Phi}}}' \quad \Rightarrow \quad \overline{\underline{\mathbf{Z}}}_R = \overline{\underline{\mathbf{\Phi}}}' \overline{\underline{\mathbf{Z}}}_L \overline{\underline{\mathbf{\Phi}}}^+ \\
\text{or}\quad &\overline{\underline{\mathbf{Y}}}_B \overline{\underline{\mathbf{\Phi}}}\, \overline{\underline{\mathbf{Y}}}_T^+ = \overline{\underline{\mathbf{\Phi}}}' \quad \Rightarrow \quad \overline{\underline{\mathbf{Y}}}_B = \overline{\underline{\mathbf{\Phi}}}' \overline{\underline{\mathbf{Y}}}_T \overline{\underline{\mathbf{\Phi}}}^+,
\end{aligned} \tag{40}
$$

the decomposition reduces to a degenerate one-sided form, and the butterflies generated by the off-side unitary matrix in the general result all reduce to pass-through operations. This is how the FFT transform first collapses layers 5–7 into layers 1–3 and then subsequently collapses layer 3 into layer 1.

A more subtle technique for reducing computational complexity involves the choice of input and output order. While it is difficult to identify any general methodology for making such a choice, judicious choice of input and/or output ordering significantly impacts the structure of associated butterfly flow graph. This is evidenced by the fact that bit reversing the order of either the inputs or the outputs in the FFT provides a fast transformation, while either eliminating the bit reversal or bit reversing both inputs and outputs eliminates the symmetries necessary to permit collapse of the flow graph.

7 Summary

Currently, MLD is more of a mathematical curiosity than a mainstream computational technique. However, the author believes that it may provide a key to a range of future advanced applications. Some of these future possibilities are considered here.

MLD provides a concise method for representing a unitary matrix in terms of its underlying independent parameters. This can be extremely useful when dealing with the effects of unitary matrices in an analytic context, where the implied relationships between the different matrix elements (which are very difficult to state directly) become important. The unitary matrix representation developed in [7], which is a central innovation enabling derivation of the principal results in [6], is a good example of the utility of such structural understanding; MLD can potentially fill a similar role.

An obvious application involves the development of various forms of "designer" fast transforms. MLD can expose other unitary matrices possessing a hidden fast computational structure. Perhaps more interesting, by imposition of the same kind of symmetry rules that lead to the FFT, it enables the possibility of starting with from a unitary matrix that is not inherently fast, and then developing an approximation that is both unitary and computationally efficient. Indeed, it should be possible to directly control trades between accuracy of the approximation and speed of computation in unique and useful ways.

An analysis of existing wavelet transforms with MLD will likely yield additional insight into their underlying structure, and may very well provide avenues for further advancement and generalization of this ubiquitously employed class of computational methods.

A related insight involves possible approaches for recapturing the orthogonality of discretely sampled variants of continuous orthogonal function sets. To the author's knowledge, there is no known general method of discretely sampling a set of continuous orthogonal functions that is guaranteed to arrive at a discretely sampled orthogonal transformation. Further, the ability to actually achieve this is very limited; only the Fourier transform (and possibly a few select wavelet transformations) is known to support this exactly. Proper use of an extended form of MLD (including a central scaling layer to address non-unitary matrices) opens the possibility of recovering discrete orthogonal transformations that closely approximate existing continuous transformations.

Two more visionary possibilities also arise. The fact that the product of two unitary matrices remains unitary implies that there must be some way to collapse the sequential concatenation of two such generalized butterfly structures from Fig. 1 (the flow graph of the product) back down to a single equivalent (the flow graph of the resulting unitary matrix). The actual computational rules for accomplishing this are currently unknown to the author, but their explicit delineation could lead to a mathematics for synthetic manipulation and multiplication of matrices, potentially with computational and/or storage advantages over standard matrix multiplication rules. Such results can usefully be thought of as developing the methods for manipulating matrices when stated in generalized polar form.

Equivalently, it is possible that MLD holds the key to the development of a method for calculating the SVD that is based around matrix decimation constructs. While it is not guaranteed that such an algorithm exists, it is reasonably clear that, if it exists, it must exploit the inherent structure of the two singular vector matrices that is exposed by MLD.

Acknowledgments This research was partially supported by the Office of Naval Research Code 321MS, via contract through Oak Ridge National Laboratory. Oak Ridge National Laboratory is managed by UT-Battelle, LLC for the United States Department of Energy under Contract DE-AC05-00OR22725.

References

1. A. Oppenheim, R. Schafer, *Digital Signal Processing* (Prentice-Hall, Englewood Cliffs, NJ, 1975), pp. 290–309 Sections 6.2 and 6.3
2. J. Polcari, Butterfly decompositions for arbitrary unitary matrices, Working Paper, Rev 2 (Feb 2014), https://drive.google.com/file/d/0B1JunjKxazANLUhiaVhiejB5eDQ
3. Wikipedia, *Unitary matrix* [Online] (2018), http://en.wikipedia.org/wiki/Unitary_matrix
4. Wikipedia, *Givens rotation* [Online] (2018), http://en.wikipedia.org/wiki/Givens_rotation
5. J. Polcari, MLD MATLAB Code Set (Oct 2017), https://drive.google.com/drive/folders/0B1JunjKxazANYmlDQkZoWWRxVVU
6. J. Polcari, Jacobian determinants of the SVD, Working Paper, Rev 3 (Sept 2017), https://drive.google.com/file/d/0B1JunjKxazANYXgzbXVyMlN3OGM
7. J. Polcari, Representing unitary matrices by independent parameters, Working Paper, Rev 0 (Oct 2016), https://drive.google.com/file/d/0B1JunjKxazANM0RtbzhGTi1ZSVU

Development of an Aft Boundary Condition for a Horizontally Towed Flexible Cylinder

Anthony A. Ruffa

1 Introduction

In the linear regime, the transverse displacements along a neutrally buoyant flexible cylinder with a tow point at $x = 0$ are governed by the following equation [1]:

$$\left[T(x) - \rho_0 \pi a_A^2 U^2\right] \frac{\partial^2 Y}{\partial x^2} - 2\rho_0 \pi a_A^2 U \frac{\partial^2 Y}{\partial x \partial t} - 2\rho_0 \pi a_A^2 \frac{\partial^2 Y}{\partial t^2} = \rho_0 \pi a_A U c_N \frac{\partial Y}{\partial t} + \rho_0 \pi a_A U^2 c_N \frac{\partial Y}{\partial x}. \tag{1}$$

Here, Y is the transverse displacement, x is the distance along the cylinder axis, $T(x) = \rho_0 \pi a_A U^2 c_T (L - x)$ is the tension, ρ_0 is the density of water, c_T and c_N are the tangential and normal hydrodynamic drag coefficients, respectively, U is the steady-state tow speed, a_A is the cylinder radius, and L is the cylinder length. Equation (1) neglects bending terms. This approximation is valid along the entire towed cylinder, except for a small region near $x = L$, i.e., the free end (where the tension approaches zero).

The numerical solution of (1) is straightforward, except for the aft boundary condition. There are numerous approaches in the literature, e.g., specifying a zero tension and other conditions at the free end [2], assuming that the cylinder is straight in a region near the free end [3], or imposing a zero bending moment at the free end (along with an equation developed from the transverse momentum balance), and employing the method of matched asymptotic expansions [4].

A. A. Ruffa (✉)
Naval Undersea Warfare Center, Newport, RI, USA
e-mail: anthony.ruffa@navy.mil

A. A. Ruffa, B. Toni (eds.), *Advanced Research in Naval Engineering*, STEAM-H: Science, Technology, Engineering, Agriculture, Mathematics & Health, https://doi.org/10.1007/978-3-319-95117-1_4

Kuo and Ruffa [5] avoid the boundary condition at the free end altogether by using a method of characteristics approach that supports a boundary condition at the critical point (i.e., where the tension and the hydrodynamic forces balance). The critical point is located significantly upstream of the free end, and the boundary condition makes use of the slopes of the two characteristics at that point. That approach avoids the region where the bending terms become important, and it removes the problem of modeling the instabilities that have been observed in the region near the free end [6]. However, the method of characteristics has not been widely adopted for this problem. Here, that boundary condition is extended to more general numerical methods, e.g., the finite difference method. Extending it to the finite element method is not presented here but would involve a similar approach.

2 First-Order Numerical Implementation

In a method of characteristics solution, one of the characteristic curves (or characteristics) approaches zero at a critical point [5] located at $x = x_c$ where

$$x_c = L - \frac{a_A}{c_T}. \tag{2}$$

Kuo and Ruffa [5] developed an aft boundary condition on the cylinder at $x = x_c$ instead of at the actual free end. The finite difference approximation to (1) for $0 \le x \le x_c$ discretizes the continuous cylinder into $n - 1$ sections via n finite difference nodes. Assuming a harmonic time dependence, i.e., $Y(x, t) = y(x)e^{i\omega t}$ (where $\omega = 2\pi f$ and f is the frequency), the resulting equation at node j (where $2 \le j \le n - 1$) is as follows:

$$\begin{aligned}
&\left[T(x_j) - \rho_0 \pi a_A^2 U^2\right] \frac{y_{j+1} - 2y_j + y_{j-1}}{\Delta x^2} \\
&\quad - 2i\omega\rho_0\pi a_A^2 U \frac{y_{j+1} - y_{j-1}}{2\Delta x} + 2\omega^2 \rho_0 \pi a_A^2 y_j \\
&\quad = i\omega\rho_0\pi a_A U c_N y_j + \rho_0 \pi a_A U^2 c_N \frac{y_{j+1} - y_{j-1}}{2\Delta x}.
\end{aligned} \tag{3}$$

Here, $x_j = (j - 1)\Delta x$, $\Delta x = x_c/(n - 1)$, and $T(x_j) = \rho_0 \pi a_A U^2 c_T (L - x_j)$. The boundary condition at $j = 1$ (i.e., at $x = 0$) is $y_1 = 1$. The boundary condition at $j = n$ (i.e., at $x = x_c$) is developed using the method of characteristics. The characteristics F_a and F_b corresponding to (1) are as follows [5]:

$$\frac{dx}{dt} = (F_a, F_b) = \frac{U}{2} \pm \left[\frac{T}{2m} - \frac{U^2}{4}\right]^{1/2}. \tag{4}$$

At $x = x_c$, $T = mU^2$, so that $F_a = U$ and $F_b = 0$ in (4). Along the characteristics, (1) reduces to

$$dU' - [(F_a, F_b) - U]\, d\Phi' + \frac{1}{2m} F'_y dt = 0. \tag{5}$$

where

$$U' = -\frac{\partial Y}{\partial t}; \tag{6}$$

$$\Phi' = -\frac{\partial Y}{\partial x}; \tag{7}$$

$$m = \pi \rho_0 a_A^2; \tag{8}$$

and

$$F'_y = \rho_0 \pi c_N a_A U \left[U' + U\Phi'\right]. \tag{9}$$

Figure 1 illustrates the characteristics F_a and F_b on the finite difference grid in the $x - t$ plane at $x = x_c$.

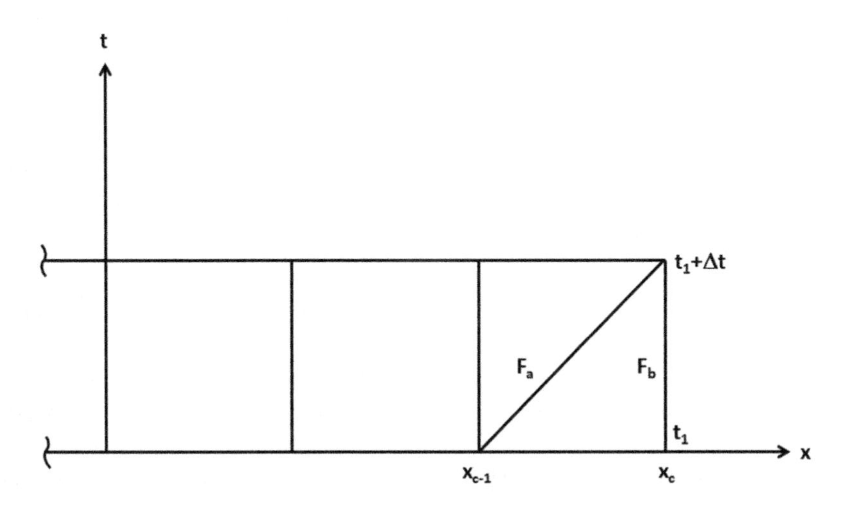

Fig. 1 The finite difference grid in the $x - t$ plane illustrating the characteristics F_a and F_b at $x = x_c$

Since F_b is vertical in the $x - t$ plane at $x = x_c$, (5) leads to an equation for $U(x_c, t)$, which effectively supports a boundary condition at $x = x_c$. Setting the time step to $\Delta t = \Delta x / F_a(n)$ ensures that F_a will intersect a finite difference node at $x = x_{c-1} = (n-1)\Delta x$. The equation at $x = x_c$ along F_a can then be approximated as follows:

$$U'(x_c, t_1 + \Delta t) - U'(x_{c-1}, t_1) + \frac{c_N U \Delta t}{2a_A}\left[U'(x_c, t_1 + \Delta t) + U\Phi'(x_c, t_1 + \Delta t)\right] = 0. \tag{10}$$

Applying a harmonic time dependence (i.e., $U'(x,t) = u'(x)e^{i\omega t}$ and $\Phi'(x,t) = \phi'(x)e^{i\omega t}$) leads to:

$$u'(x_c) - u'(x_{c-1})e^{-i\omega\Delta t} + \frac{c_N U \Delta t}{2a_A}\left[u'(x_c) + U\phi'(x_c)\right] = 0. \tag{11}$$

Evaluating the temporal derivatives in (5) analytically along F_b leads to

$$i\omega u'(x_c) + i\omega U\phi'(x_c) + \frac{c_N U}{2a_A}\left[U\phi'(x_c) + u'(x_c)\right] = 0, \tag{12}$$

or

$$u'(x_c) + U\phi'(x_c) = 0. \tag{13}$$

Finally, substituting (13) into (11) leads to:

$$u'(x_c)e^{i\omega\Delta t} - u'(x_{c-1}) = 0. \tag{14}$$

From (6), it follows that

$$y(x_c)e^{i\omega\Delta t} - y(x_{c-1}) = 0. \tag{15}$$

Equation (13) indicates that $F_y' = 0$ and $dy = 0$ along the characteristic F_a intersecting the finite difference grid at $x = x_c$. Equation (15) is the boundary condition used at $x = x_c$. The results are shown in Fig. 2 for the following parameters: $U = 10$ knots, $c_T = 0.0025$, $c_N = c_T/2$, $L = 100$ m, $a_A = 1.5$ in., and $f = 0.1$ Hz.

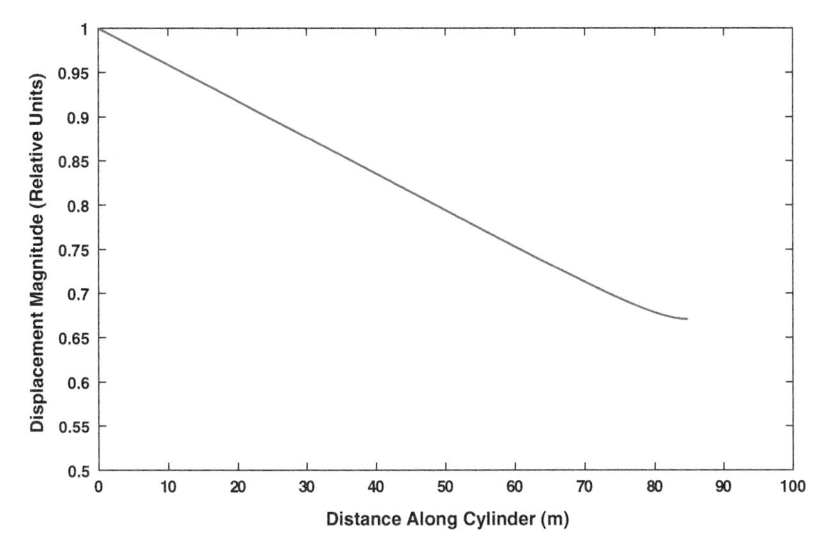

Fig. 2 The magnitude of the displacement vector y. Here, $L = 100\,\text{m}$ and $x_c = 84.76\,\text{m}$. The remainder of the cylinder (i.e., $x_c \leq x \leq L$) is not modeled

3 Second-Order Numerical Implementation

The derivative estimated along F_a in (10) is not consistent with a second-order finite difference scheme, which requires three points along each characteristic (Fig. 3) to estimate the derivative. Ensuring that F_a intersects finite difference nodes at $x = x_{c-1}$ and $x = x_{c-2}$ requires two different discretization lengths, i.e., h_1 and h_2, where

$$h_1 \cong \Delta t \sqrt{1 + F_a(n)^2}; \tag{16}$$

and

$$h_2 \cong \Delta t \sqrt{1 + F_a(n-1)^2}. \tag{17}$$

A backward difference formula with unequal spacing [7] can approximate the second-order derivative along the characteristic F_a, i.e.,

$$\begin{aligned}\frac{dU'}{ds} = &\frac{2h_1 + h_2}{h_1(h_1 + h_2)} U'(x_c, t_1 + 2\Delta t) - \frac{h_1 + h_2}{h_1 h_2} U'(x_{c-1}, t_1 + \Delta t) \\ &+ \frac{h_1}{(h_1 + h_2)h_2} U'(x_{c-2}, t_1).\end{aligned} \tag{18}$$

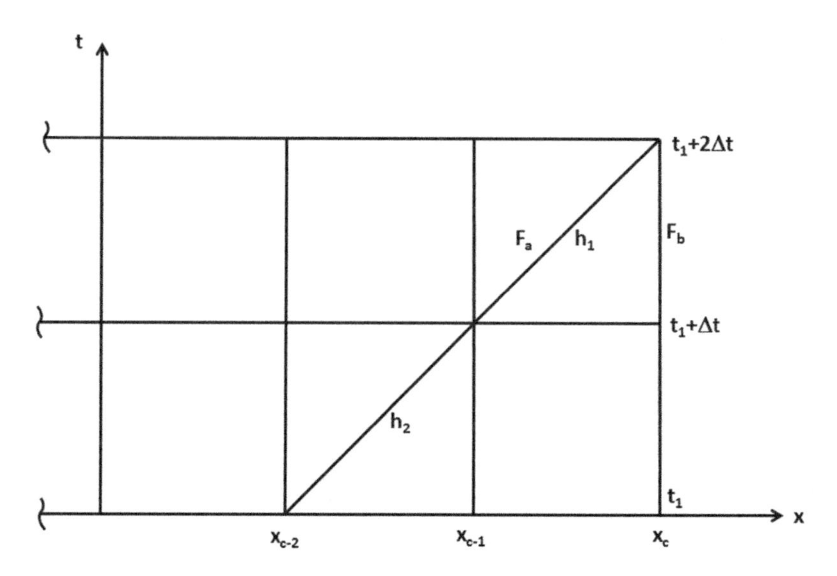

Fig. 3 The numerical grid in the $x-t$ plane showing the characteristics F_a and F_b at $x = x_c$. Note that $x_c - x_{c-1} \neq x_{c-1} - x_{c-2}$ because of the variation of F_a as a function of x

Applying a harmonic time dependence and noting that $F_y' = 0$ at $x = x_c$ as before (so that $dy = 0$ along F_a) leads to the boundary condition, i.e.,

$$\frac{2h_1 + h_2}{h_1(h_1+h_2)} y(x_c) e^{2i\omega\Delta t} - \frac{h_1 + h_2}{h_1 h_2} y(x_{c-1}) e^{i\omega\Delta t} + \frac{h_1}{(h_1+h_2)h_2} y(x_{c-2}) = 0. \tag{19}$$

For the second-order finite difference approximation scheme, $\Delta x_1 = h_2/\Delta t = \eta_1$ over the entire solution domain, except between x_{c-1} and x_c, where $\Delta x_2 = h_1/\Delta t = \eta_2$. As a result, there are *two* sets of finite difference equations: Eq. (3) for $2 \leq j \leq n-2$, and for $j = n-1$, the equation is given below:

$$\begin{aligned} &2\left[T(x_j) - \rho_0 \pi a_A^2 U^2\right] \frac{\eta_2 y_{j+1} - (\eta_1 + \eta_2) y_j + \eta_1 y_{j-1}}{\eta_1 \eta_2 (\eta_1 + \eta_2)} + 2\omega^2 \rho_0 \pi a_A^2 y_j \\ &\quad - i\omega \rho_0 \pi a_A U c_N y_j - \rho_0 \pi a_A U \left(2i\omega a_A + U c_N\right) \\ &\quad \times \left[\frac{\eta_2^2 y_{j+1} + (\eta_1^2 - \eta_2^2) y_j - \eta_1^2 y_{j-1}}{\eta_1 \eta_2 (\eta_1 + \eta_2)}\right] = 0. \end{aligned} \tag{20}$$

Figure 4 shows the result using (19) as a boundary condition, and Fig. 5 shows the magnitude of the difference between the first- and second-order approximations, indicating that the first-order truncation error along F_a is small. The variation in F_a is small at $x \approx x_c$, i.e., $F_a(n)/F_a(n-1) \cong 0.9986$.

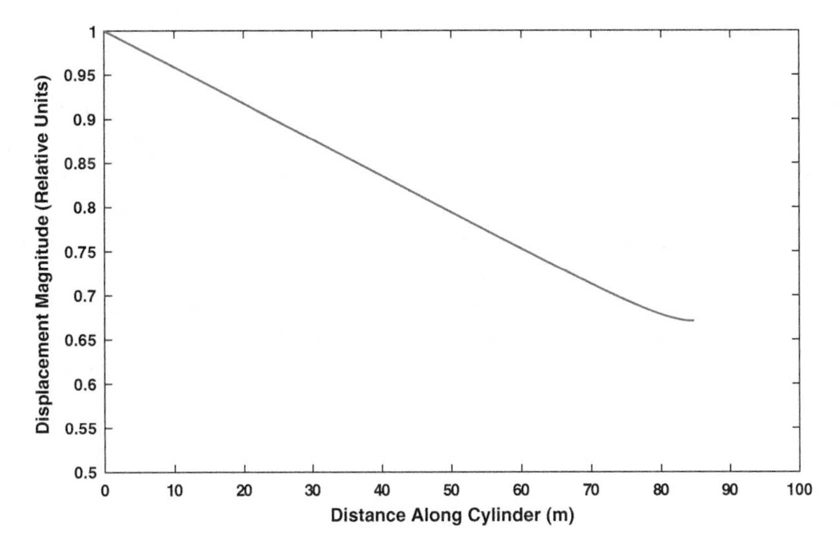

Fig. 4 The magnitude of the displacement vector y for the second-order solution

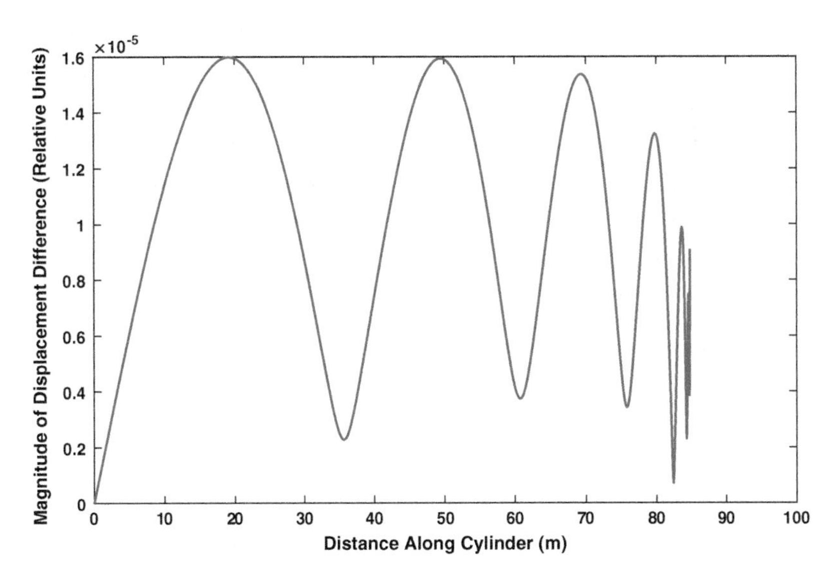

Fig. 5 The magnitude of the difference in the displacement vectors for the first-order and second-order solutions

4 Conclusions

The method of characteristics applied at the critical point on a towed flexible cylinder led to an aft boundary condition for a finite difference approach. The boundary condition consisted of the equation $dy = 0$ (where y is the magnitude of the transverse displacement) along the nonzero characteristic intersecting the finite difference grid at $x = x_c$. That boundary condition can also be applied to finite element approaches, and even to the nonlinear evolution of a towed system during more general maneuvers. For example, Schram and Reyle [8] used the method of characteristics to simulate the three-dimensional motion of a cable-body system. This approach can support an aft boundary condition at the critical point, even when the remainder of the system is modeled with finite differences or finite elements.

References

1. M.P. Païdoussis, Dynamics of cylindrical structures subjected to axial flow. J. Sound Vib. **29**, 365–385 (1973)
2. V.K. Srivastava, Y.V.S.S. Sanyasiraju, M. Tamsir, Dynamic behavior of underwater towed-cable in linear profile. Int. J. Sci. Eng. Res. **2**(7), 1–10 (2011)
3. C.M. Ablow, S. Schechter, Numerical simulation of undersea cable dynamics. Ocean Eng. **10**(6), 443–457 (1983)
4. A.P. Dowling, The dynamics of towed flexible cylinders, Part 1. Neutrally buoyant elements. J. Fluid Mech. **187**, 507–532 (1988)
5. E.Y.T. Kuo, A.A. Ruffa, A numerical treatment of a free end termination for a horizontally towed flexible cylinder. Technical Memorandum No. 871114, Naval Underwater Systems Center, New London, CT (1987)
6. E. De Langre, M.P. Païdoussis, O. Doaré, Y. Modarres-Sadeghi, Flutter of long flexible cylinders in axial flow. J. Fluid Mech. **571**, 371–389 (2007)
7. A.K. Singh, B.S. Bhadauria, Finite difference formulae for unequal sub-intervals using Lagrange's interpolation formula. Int. J. Math. Anal. **3**, 815–827 (2009)
8. J.W. Schram, S.P. Reyle, A three-dimensional dynamic analysis of a towed system. J. Hydronaut. **2**, 213–220 (1968)

Tracking with Deterministic Batch Trackers

Steven Schoenecker

1 Deterministic Batch Trackers

In this chapter, we develop two deterministic batch trackers—the Maximum Likelihood Probabilistic Data Association (ML-PDA) tracker and the Maximum Likelihood Probabilistic Multi-Hypothesis Tracker (ML-PMHT). Both are non-Bayesian trackers that assume a target (that is to be tracked) has some unknown deterministic motion that is corrupted by measurement noise. In contrast, in the Bayesian model, the truth itself is assumed to contain uncertainty (i.e., process noise). Additionally, ML-PDA and ML-PMHT are usually implemented as batch trackers, where a set of scans or updates over some period of time is processed concurrently. In contrast, an iterative/recursive tracker contains probabilistic knowledge of the current state (in the form of a prior probability distribution), and then automatically updates this knowledge with every new scan.

For deterministic trackers, the underlying truth can be "parameterized" mathematically. For instance, for a given batch of data that covers a relatively short span of time, it is often assumed that if a target is present, it is moving in a straight line. Such a straight line can be parameterized by position and velocity at the start of the batch; given this target state, and the time elapsed since the start of the batch, the target position can be calculated for any point in the batch.

In theory, a batch could cover a very extended amount of time; however, this would require the motion parameterization for the target to be accurate over that length of time. In reality, this is not often done, because it is difficult to parameterize a target's motion for extended periods (for instance, targets maneuver). In reality,

S. Schoenecker (✉)
Naval Undersea Warfare Center, Newport, RI, USA
e-mail: steven.schoenecker@navy.mil

A. A. Ruffa, B. Toni (eds.), *Advanced Research in Naval Engineering*, STEAM-H: Science, Technology, Engineering, Agriculture, Mathematics & Health, https://doi.org/10.1007/978-3-319-95117-1_5

batch lengths tend to be relatively short (seconds or minutes, depending on the target type). This way, parameterizing the target motion during that short period of time can be done to a reasonable accuracy with a simple target motion parameterization (again, usually a straight line, although other parameterizations do exist).

1.1 Maximum Likelihood Probabilistic Data Association (ML-PDA) tracker

We first present a tracker/estimator that is relatively "common" in the literature, the Maximum Likelihood Probabilistic Data Association (ML-PDA) tracker. This algorithm was initially developed in [1] for a passive sonar application. In [2] it was modified to work for an electro-optical sensor tracking an aircraft. Finally, in [3, 4], and [5] it was extended to a work in a multistatic active sonar framework. As with all maximum likelihood (ML) techniques, a log-likelihood ratio (LLR) is formulated, and the state that yields the maximum log-likelihood value is considered as the potential target state. The ML-PDA LLR is constructed by making assumptions about a single target with deterministic, parameterizable motion in the presence of clutter. The assumptions are [6]

- A single target is present in each frame with known detection probability P_d. Detections are independent across frames.
- There are zero or one measurements per frame from the target.
- The kinematics of the target are deterministic. The motion is usually parameterized as a straight line, although any other parameterization is valid.
- False detections are uniformly distributed in the search volume.
- The number of false detections is Poisson distributed with known clutter density.
- Amplitudes of target and false detections are Rayleigh distributed. The parameter of each Rayleigh distribution is known.
- Target measurements are corrupted with zero-mean Gaussian noise.
- Measurements at different times, conditioned on the parameterized state, are independent.

Using these assumptions, the LLR for ML-PDA is constructed by first defining the following mutually exclusive and exhaustive events:

ϵ_0	The target is not detected; all measurements are clutter-originated.
$\epsilon_j \quad (j = 1, \ldots, m_i)$	The target is detected and the detection is from the jth measurement; all other measurements are clutter-originated.

Using these events, the likelihood ratio (LR) for a scan of data $\phi[Z(i), \mathbf{x}]$ is written as

$$\phi[Z(i), \mathbf{x}] = \frac{p[Z(i), m_i, \boldsymbol{\varepsilon}|\mathbf{x}]}{p[Z(i), m_i, \varepsilon_0|\mathbf{x}]} \tag{1}$$

The numerator is the joint (mixed) probability density function/probability mass function (PDF/PMF) of all the measurements in the ith frame, the number of measurements, and the set of *all* events $\boldsymbol{\varepsilon}$. The numerator can be expanded in the following manner using the total probability theorem:

$$p[Z(i), m_i, \boldsymbol{\varepsilon}|\mathbf{x}] = p[Z(i), m_i|\varepsilon_0]p(\varepsilon_0) + \sum_{j=1}^{m_i} p[Z(i)|m_i, \varepsilon_j, \mathbf{x}]p(m_i, \varepsilon_j|\mathbf{x}) \quad (2)$$

For the first term on the right-hand side in this expression, the conditioning on $\mathbf{x}$ has been dropped since this term accounts for all measurements being clutter-originated and thus is not dependent on the target state. The probability of event ε_0 occurring is just the probability of the target not being detected; this is simply given by

$$p(\varepsilon_0) = 1 - P_d \quad (3)$$

The joint PDF/PMF of the measurements in the ith scan and the number of measurements, conditioned on ε_0, can be simply re-written as

$$p[Z(i), m_i|\varepsilon_0] = p[Z(i)|m_i, \varepsilon_0]p(m_i|\varepsilon_0) \quad (4)$$

The PDF of the measurements conditioned on the number of measurements m_i and the event they are all clutter-originated can be written by considering two of the ML-PDA assumptions—clutter measurements are uniformly distributed in the search volume, and amplitudes of false detections are Rayleigh distributed.

$$p[Z(i)|m_i, \varepsilon_0] = \frac{1}{V^{m_i}} \prod_{j=1}^{m_i} p_c^{\tau}(a_j) \quad (5)$$

The notation $p_c^{\tau}(a_j)$ denotes a thresholded Rayleigh distribution for clutter measurements—this (as well as the target amplitude distribution) will be discussed shortly.

The probability mass function (PMF) for the number of measurements conditioned on the event that all measurements are clutter-originated is determined from another of the ML-PDA assumptions—namely that the number of clutter measurements is Poisson distributed. This PMF is expressed as

$$p(m_i|\varepsilon_0) = \frac{e^{-\lambda V}(\lambda V)^{m_i}}{m_i!} \quad (6)$$

Here, λ is the clutter density, so the product λV is the expected number of clutter measurements. Taking the product of (3), (5), and (6) gives

$$p[Z(i), m_i, |\varepsilon_0] = (1 - P_d)\frac{1}{V^{m_i}} \prod_{j=1}^{m_i} p_c^{\tau}(a_j)\frac{e^{-\lambda V}(\lambda V)^{m_i}}{m_i!} \quad (7)$$

Now, turning to the second term in the right-hand side of (2), the joint PDF for a batch of measurements conditioned on m_i measurements, event ε_j, and $\mathbf{x}$ is written as

$$p[Z(i)|m_i, \varepsilon_j, \mathbf{x}] = \frac{1}{V^{m_i-1}} \prod_{k=1,k\neq j}^{m_i} p_c^{\tau}(a_k) p[z(j)|\mathbf{x}] p_t^{\tau}(a_j) \tag{8}$$

Here, $p[z(j)|\mathbf{x}]$ is a target-centered Gaussian (another one of the ML-PDA assumptions), and $p_t^{\tau}(a_j)$ is a thresholded Rayleigh *target* amplitude distribution. Note that for this event (ε_j), the jth measurement is from the target and the other $m_i - 1$ measurements are from clutter. The joint PMF of the number of measurements and event ε_j conditioned on $\mathbf{x}$ can be simplified to

$$p(m_i, \varepsilon_j|\mathbf{x}) = p(m_i|\varepsilon_j) p(\varepsilon_j) \tag{9}$$

This simplification is possible since this joint PMF is independent of $\mathbf{x}$. The PMF for m_i again follows a Poisson distribution

$$p(m_i|\varepsilon_j) = \frac{e^{-\lambda V}(\lambda V)^{m_i-1}}{(m_i - 1)!} \tag{10}$$

Note that this Poisson distribution is slightly different than that of the clutter-only case for event ε_0 (6), since for all other events, one of the measurements is target-originated.

The PMF for ε_j, the probability of the target being detected due to measurement j is

$$p(\varepsilon_j) = P_d \frac{1}{m_i} \tag{11}$$

The term $1/m_i$ is due to the assumption that *any* of the m_i measurements are equally likely to be the measurement originating from the target. Thus, the summation term in (2) can be written as

$$\sum_{j=1}^{m_i} p[Z(i)|m_i, \varepsilon_j, \mathbf{x}] p(m_i, \varepsilon_j|\mathbf{x}) =$$

$$\sum_{j=1}^{m_i} \frac{P_d}{V^{m_i-1}} \frac{e^{-\lambda V}(\lambda V)^{m_i-1}}{(m_i - 1)!} \frac{1}{m_i} \prod_{k=1,k\neq j}^{m_i} p_c^{\tau}(a_k) p[z(j)|\mathbf{x}] p_t^{\tau}(a_j) \tag{12}$$

Defining the amplitude likelihood ratio as

$$\rho_a(a_j) = \frac{p_t^\tau(a_j)}{p_c^\tau(a_j)} \tag{13}$$

and multiplying the top and bottom of (12) by $p_c^\tau(a_j)$ allows for the following simplification:

$$\sum_{j=1}^{m_i} p[Z(i)|m_i, \varepsilon_j, \mathbf{x}]p(m_i, \varepsilon_j|\mathbf{x}) =$$

$$\sum_{j=1}^{m_i} \frac{P_d}{V^{m_i-1}} \frac{e^{-\lambda V}(\lambda V)^{m_i-1}}{(m_i-1)!} \frac{1}{m_i} \prod_{k=1}^{m_i} p_c^\tau(a_k) p[z(j)|\mathbf{x}]\rho_a(a_j) \tag{14}$$

Combining (7) and (14), recognizing that the denominator in (1) is just (7) without the term $1 - P_d$, and simplifying gives the expression for the ML-PDA LR for a single scan of data:

$$\phi[Z(i), \mathbf{x}] = 1 - P_d + \frac{P_d}{\lambda} \sum_{j=1}^{m_i} p[z(j)|\mathbf{x}]\rho_a(a_j) \tag{15}$$

The ML-PDA algorithm is a batch tracker; assuming there are N_w scans, and taking the logarithm to produce the log-likelihood ratio produces the final result:

$$\Lambda(\mathbf{x}, Z) = \sum_{i=1}^{N_w} \log \left\{ 1 - P_d + \frac{P_d}{\lambda} \sum_{j=1}^{m_i} p[z(j)|\mathbf{x}]\rho_a(a_j) \right\} \tag{16}$$

Finally, the constant $N_w \log(1 - P_d)$ is sometimes subtracted from (16) to produce a modified ML-PDA LLR

$$\Lambda_{mod}(\mathbf{x}, Z) = \sum_{i=1}^{N_w} \log \left\{ 1 + \frac{P_d}{\lambda(1 - P_d)} \sum_{j=1}^{m_i} p[z(j)|\mathbf{x}]\rho_a(a_j) \right\} \tag{17}$$

It is possible to do this because ML-PDA is a maximum-likelihood technique; subtracting off a constant from the entire function will not change the location of the target state $\mathbf{x}$ that produces the maximum likelihood value, i.e.,

$$\hat{\mathbf{x}} = \arg\max_{\mathbf{x}} \Lambda(\mathbf{x}, Z) \tag{18}$$

The formulation in (17) has the advantage of having the floor of the LLR remain at zero regardless of the batch length.

Finally, we note that it is possible to generate the ML-PDA LLR by assuming that the number of measurements in a scan (m_i) is given. (In the above derivation, this was not assumed.) However, if this is assumed, then it is not necessary to invoke the Poisson PMF in the above derivation. The final result in this case turns out to be exactly the same as that presented in (17).

1.2 ML-PMHT

We now develop the LLR for the second deterministic batch tracker presented in this chapter, the Maximum Likelihood Probabilistic Multi-Hypothesis Tracker (ML-PMHT). This algorithm is not as common in the literature, but it has definite advantages over ML-PDA. First, it is very easy to formulate the ML-PMHT LLR for multiple targets. It is in theory possible to formulate the ML-PDA LLR for an arbitrary number of targets as well, but for any more than two or three targets, this becomes extremely combinatorically complex. Additionally, it is possible to compute the Crámer-Rao Lower Bound (CRLB) for ML-PMHT real time; it is not possible to do this for ML-PDA. This (in addition to the fact that ML-PMHT can be shown to be an efficient estimator) allows this CRLB to be used to approximate the state covariance on the estimates the algorithm produces. These two conditions make the ML-PMHT more appealing than ML-PDA.

The ML-PMHT algorithm was initially developed in [7] and subsequently expanded in [8, 9], and [10]. The assumptions used to develop its LLR are very similar to those of ML-PDA, with the following major difference: whereas ML-PDA assumes that only zero or one measurement could be target-originated in a single scan, the ML-PMHT algorithm assumes that *any* number of measurement in a scan (up to all of them) can originate from the target. Because of this, the ML-PMHT LLR is *much* easier to develop.

To this end, define the event set for a single measurement:

ϵ_0	The measurement is clutter-originated.
ϵ_1	The measurement is target-originated.

The LR $\phi_1[z(j), \mathbf{x}]$ for a single measurement can be written as

$$\phi_1[z(j), \mathbf{x}] = \frac{p[z(j), \boldsymbol{\varepsilon}|\mathbf{x}]}{p[z(j)|\varepsilon_0, \mathbf{x}]} \tag{19}$$

Again, according to the total probability theorem, this can be expanded to

$$\phi_1[z(j), \mathbf{x}] = \frac{p[z(j)|\varepsilon_0]p(\varepsilon_0) + p[z(j)|\varepsilon_1, \mathbf{x}]p(\varepsilon_1)}{p[z(j)|\varepsilon_0]} \tag{20}$$

Note that again we have dropped the conditioning on $\mathbf{x}$ where the parameters in question and $\mathbf{x}$ are independent. In the numerator, the first term can be written as

$$p[z(j)|\varepsilon_0]p(\varepsilon_0) = \frac{\pi_0}{V} p_c^\tau(a_j) \tag{21}$$

where π_0 is defined as the probability of a measurement originating from clutter. The second term is written as

$$p[z(j)|\varepsilon_1, \mathbf{x}]p(\varepsilon_1) = \pi_1 p[z(j)|\mathbf{x}]p_t^\tau(a_j) \tag{22}$$

where π_1 is defined as the probability of a measurement originating from the target (i.e., $p(\varepsilon_1)$). Putting these together and adding the denominator results in

$$\phi_1[\mathbf{x}, z(j)] = \frac{\frac{\pi_0}{V} p_c^\tau(a_j) + \pi_1 p[z(j)|\mathbf{x}]p_t^\tau(a_j)}{\frac{1}{V} p_c^\tau(a_j)} \tag{23}$$

Simplifying, and taking advantage of the fact that multiple measurements in a scan are allowed to be target originated leads to the LR for a batch of measurements

$$\phi(\mathbf{x}, Z) = \prod_{i=1}^{N_w} \prod_{j=1}^{m_i} \left\{\pi_0 + V\pi_1 p[z(j)|\mathbf{x}]\rho_a(a_j)\right\} \tag{24}$$

Taking the logarithm simply turns the products into sums

$$\Lambda(\mathbf{x}, Z) = \sum_{i=1}^{N_w} \sum_{j=1}^{m_i} \log\left\{\pi_0 + V\pi_1 p[z(j)|\mathbf{x}]\rho_a(a_j)\right\} \tag{25}$$

Finally, again subtracting out a constant (in this case $\sum_{i=1}^{N_w} \sum_{j=1}^{m_i} \log(\pi_0)$) so the floor of the LLR is zero leaves the expression in the final form:

$$\Lambda_{mod}(\mathbf{x}, Z) = \sum_{i=1}^{N_w} \sum_{j=1}^{m_i} \log\left\{1 + \frac{\pi_1 V}{\pi_0} p[z(j)|\mathbf{x}]\rho_a(a_j)\right\} \tag{26}$$

Again, ML-PMHT is a maximum likelihood technique; the state vector $\mathbf{x}$ that maximizes $\Lambda_{mod}(\mathbf{x}, Z)$ over a batch of measurements is selected as the most likely target solution.

2 Amplitude Ratio

In this section we briefly discuss two common amplitude ratio models for $\rho_a(a_j)$ that are found in the literature. The first is a "Gaussian fluctuating model"—this comes about by assuming that the distribution of both the target-generated amplitudes and the clutter-generated amplitudes end up being Rayleigh distributed (this model was part of the "original" ML-PDA assumptions). Under the second scenario, the clutter is assumed to be "heavier-tailed." Such heavy-tailed clutter can be described by various different probability density functions; we briefly describe one, the K-distribution, that is popular in the literature.

First, it is assumed there exists some real, continuous-valued, continuous-time, band-limited signal $s(t)$ that is received by an active sonar system. Such a system will typically calculate the analytic signal $s_a(t)$ by applying a Hilbert transform to $s(t)$. This "removes" the negative frequency component (in the process giving $s_a(t)$ an imaginary part) and then calculating a basebanded signal $s_{bb}(t)$ by demodulating out the carrier frequency f_c (the center frequency of the real band-limited signal $s(t)$).[1] At this point, the complex basebanded signal is sampled via analog-to-digital conversion; this sampled signal is denoted $s[j]$ (again, note that $s[j]$ has both real and imaginary parts). It can be written as

$$s[j] = a[j]e^{i\theta} \tag{27}$$

The *amplitude* $a[j]$ of $s[j]$ is just calculated as

$$a[j] = \left(s[j]s[j]^H\right)^{\frac{1}{2}} \tag{28}$$

The *intensity* of $\zeta[j]$ is just the square of the amplitude

$$\zeta[j] = s[j]s[j]^H \tag{29}$$

The remainder of this chapter will explore the distributions typically used in tracking algorithms for $a[j]$ and $\zeta[j]$ when produced by a target or by clutter.

2.1 Fluctuating Gaussian Model

The fluctuating Gaussian model, sometimes called a *Swerling I* model [11, 12], comes about by assuming that the measurement-originating mechanism (i.e., either the target or a clutter generator) consists of multiple, similarly sized objects. Their

[1] There are several different ways of actually computing $s_{bb}(t)$; however, they will all produce a complex signal with a spectrum centered at $f = 0$.

size is similar enough that all the scatterers are assumed to have relatively equal contributions to $s(t)$, and there are enough of them that the Central Limit Theorem can be applied. Because of this, the assumption is made that envelope of $s[j]$ has a complex zero-mean Gaussian distribution [13]. If $s[j]$ is target originated (or "signal"), the real and imaginary components will each have variance $\sigma_t^2/2$. If $s[j]$ is clutter-originated (or "noise"), the variance of the components will be $\sigma_c^2/2$. Since the Gaussian distribution is closed under addition, the signal-plus-noise distribution will be zero-mean complex Gaussian as well; the real and imaginary components will have variance $\sigma^2/2 = \sigma_t^2/2 + \sigma_c^2/2$. From this, it is straightforward to derive the PDF for intensity and amplitude. The real and imaginary Gaussian components are assumed to be independent; the intensity ζ is then seen to be the sum of the square of two zero-mean Gaussians; this is easily shown to be an exponential distribution with mean value σ^2

$$f_\zeta(\zeta) = \frac{1}{\sigma^2} e^{-\frac{\zeta}{\sigma^2}} \qquad \zeta \geq 0 \tag{30}$$

The amplitude A is just the square root of the intensity; it is again easily shown that the square root of an exponential random variable with mean σ^2 has a Rayleigh distribution given by

$$f_a(a) = \frac{2a}{\sigma^2} e^{-\frac{a^2}{\sigma^2}} \qquad a \geq 0 \tag{31}$$

From these expressions, it is easy to calculate the amplitude likelihood ratio. Without loss of generalization, it can be assumed that the noise power (the intensity) has been normalized, so the variance for the signal-plus-noise case can be expressed as

$$\sigma^2 = \frac{\sigma_c^2 + \sigma_t^2}{\sigma_c^2} = 1 + \frac{\sigma_t^2}{\sigma_c^2} \triangleq 1 + d \tag{32}$$

where d is the signal-to-noise ratio, or the deflection coefficient. The noise-only case by definition has $\sigma^2 = 1$; the PDF of the intensity in this case is

$$f_\zeta(\zeta) = e^{-\zeta} \qquad \zeta \geq 0 \tag{33}$$

The PDF of the amplitude for the noise-only case is

$$f_a(a) = 2ae^{-a^2} \qquad a \geq 0 \tag{34}$$

From these expressions the amplitude likelihood ratio $\rho_a(a)$ (or equivalently if operating with intensity, $\rho_\zeta(\zeta)$) can be calculated.

$$\rho_\zeta(\zeta) = \rho_a(a) = \frac{f_\zeta(\zeta|\text{signal plus noise})}{f_\zeta(\zeta|\text{noise})} = \frac{f_a(a|\text{signal plus noise})}{f_a(a|\text{noise})} \tag{35}$$

A typical active sonar system will threshold the matched-filter output; the intensity PDFs given by (30) and (33) now become

$$f_\zeta(\zeta\,|\text{signal plus noise}) = \frac{1}{P_d}\frac{1}{1+d}e^{-\frac{\zeta}{1+d}} \qquad \zeta \geq \tau_\zeta \tag{36}$$

and

$$f_\zeta(\zeta\,|\text{noise}) = \frac{1}{P_{fa}}e^{-\zeta} \qquad \zeta \geq \tau_\zeta \tag{37}$$

where P_d and P_{fa} are the right-tailed integrals (starting at the threshold intensity τ_ζ) of (30) and (33), respectively. Combining (35), (36), and (37) gives a final expression for $\rho_a(a_j)$ for a fluctuating Gaussian/Swerling I model:

$$\rho_\zeta(\zeta) = \frac{P_{fa}}{P_d}\frac{1}{1+d}\exp\left(\frac{\tau_\zeta d}{1+d}\right)\exp\left(\frac{\zeta d}{1+d}\right) \qquad \zeta \geq \tau_\zeta \tag{38}$$

If we are operating in amplitude space, then τ_I is replaced with τ_a^2, and ζ is replaced with a^2

$$\rho_a(a) = \frac{P_{fa}}{P_d}\frac{1}{1+d}\exp\left(\frac{\tau_a^2 d}{1+d}\right)\exp\left(\frac{a^2 d}{1+d}\right) \qquad a \geq \tau_a \tag{39}$$

2.2 *Heavy-Tailed Clutter*

The fluctuating Gaussian amplitude model developed above describes the signal-plus-noise and the noise-only amplitude distributions with a Rayleigh PDF, or the signal-plus-noise and the noise-only intensity distributions with an exponential PDF. These are "light-tailed" distributions; they come from the assumption that the mechanism producing the scattering for both clutter and target consists of many similar scatterers. The mathematical effect of this light-tailed assumption is that high-amplitude measurements are probabilistically much more likely to be target-originated; this model basically discounts the chance that any high-amplitude measurements are clutter-originated. Work in recent years has started to explore more heavy-tailed distributions to describe active sonar clutter (which makes it possible for high-amplitude measurements to be clutter-originated). Distributions explored include the Weibull distribution, the Generalized Pareto distribution, and the K-distribution [14]. Here, we briefly discuss the last of these, the K-distribution, since this has a realistic physical explanation for heavy-tailed clutter [15–19], and [20].

If instead of the clutter being due to many similar scatters, where the Central Limit Theorem can be invoked, consider clutter to be originated from a fewer number of larger scatterers. The following argument is very similar to that found in [21]. Consider the I and Q components (i.e., the real and imaginary components) of $s[j]$. In the fluctuating Gaussian/Swerling I model, each of these was modeled as a zero-mean Gaussian. Let x_r denote the in-phase (I) amplitude of $s[j]$ and x_i denote the quadrature (Q) component of the signal (we can ignore the phase of the signal without loss of generality). These components can be modeled as

$$\begin{aligned} x_r &= \sqrt{v}u_1 \\ x_i &= \sqrt{v}u_2 \end{aligned} \tag{40}$$

Here, u_1 and u_2 are zero-mean Gaussian RVs with variance $\sigma^2/2$. If v is a constant, then the quantity

$$\zeta = x_r^2 + x_i^2 = v(u_1^2 + u_2^2) \tag{41}$$

is seen to be an exponential RV with expected value $v\sigma^2$. This is precisely the intensity produced by the fluctuating Gaussian model in the previous section. However, to account for the fact that there are fewer scatters (which will lead to higher variability in ζ) let the quantity v itself be a random variable. The quantity $w = (u_1^2 + u_2^2)$ is an exponential random variable with mean σ^2. Now make the assumption that v is Gamma distributed with mean *and* variance equal to $1/\sigma^2$ (in terms of shape and scale parameters, the shape parameter $\alpha = 1/\sigma^2$ and the scale parameter $\beta = 1$). In this case, where

$$\zeta = vw \tag{42}$$

this product can be shown to have a K-intensity distribution, which is given by

$$p_\zeta(\zeta) = \frac{2}{\lambda\Gamma(\alpha)}\left(\frac{\zeta}{\lambda}\right)^{(\alpha-1)/2} K_{\alpha-1}\left(2\sqrt{\frac{\zeta}{\lambda}}\right) \qquad \zeta > 0 \tag{43}$$

The parameter α now becomes the K-distribution shape parameter, $\lambda = 1/\alpha$, and K_ν is the Basset function (a modified Bessel function of the second kind) [22]. The parameter α for this distribution determines the "heaviness" of the tail; the smaller the value of α, the heavier the tail (i.e., the more the probability mass is spread out to the right). Although not apparent from this expression, the expected value for K-distributed intensity is 1 (which is necessary to model noise that has been normalized). Note that this PDF describes K-distributed *intensity*; K-distributed *amplitude* is given by

$$p_a(a) = \frac{4}{\sqrt{\lambda}\Gamma(\alpha)}\left(\frac{a}{\sqrt{\lambda}}\right)^{\alpha} K_{\alpha-1}\left(\frac{2a}{\sqrt{\lambda}}\right) \qquad a > 0 \tag{44}$$

The expressions in (44) and (43) describe K-distributed clutter amplitudes and intensity, respectively. Unlike the fluctuating Gaussian case, where it was possible to compute a closed-form expression for the signal-plus-noise case and as a result, calculate a closed-form expression for $\rho_a(a)$ or $\rho_\zeta(\zeta)$, it is not possible to compute an analytic expression for the signal-plus-noise case with the K-distribution. Instead, a PDF for the signal-only model must be developed (e.g., a Rayleigh distribution for the amplitude of a Swerling I target). Then, the PDF for the signal-plus-noise can be numerically calculated by convolving the PDF of the signal amplitude with the PDF of the clutter amplitude (44). (Alternatively, this could be done in the intensity domain.) Then, the amplitude likelihood ratio $\rho_a(a)$ can be computed by dividing this expression by the clutter-only PDF.

3 Threshold Determination for ML-PMHT

The ML-PMHT tracker works by finding the state vector $\hat{\mathbf{x}}$ that, given a batch of data, maximizes the ML-PMHT LLR. However, there is no guarantee that this maximum point is actually caused by a target. The ML-PMHT LLR surface, for either passive or active applications, is highly non-convex. To determine whether or not the global maximum on the LLR is caused by the target or is caused by clutter, it is necessary to statistically determine the peak point in the ML-PMHT that is caused by clutter measurements only. Once this PDF is determined, it is possible to select an acceptable false alarm rate and set a tracking threshold per the Neyman-Pearson Lemma. Peak LLR values found above this threshold are accepted as targets; peak LLR values below this level are rejected as clutter.

The PDF that statistically describes the peak point in the ML-PMHT LLR is determined in stages. This involves two main subjects: random-variable transformations and extreme-value theory. To determine the PDF of the maximum value of the ML-PMHT LLR, we assume only clutter measurements are present in the ML-PMHT LLR given by (26). The first step in the process is to assume that the ML-PMHT LLR for one measurement is just a random-variable transformation. Then the entire ML-PMHT LLR (again, given by (26)) can be thought of as another RV; if there are M clutter measurements in a batch; the PDF that describes all the possible LLR values for the batch of measurements is just the single-measurement PDF convolved with itself $M - 1$ times (this is possible because one of the ML-PMHT assumptions is that all measurements, conditioned on the target state $\mathbf{x}$, are independent). Finally, we are interested in the PDF of the *peak value* in the LLR that is caused by clutter measurements; for this, we turn to extreme-value theory.

The approach described below is for ML-PMHT. It is theoretically possible to do this as well for ML-PDA; however, the form of the ML-PDA LLR makes the random variable transformations that would be necessary extremely difficult. In contrast, the form of the ML-PMHT LLR makes these transformations analytically tractable.

3.1 Single Measurement Random-Variable Transformation

To determine the threshold,[2] start with the ML-PMHT assumption that clutter measurements are uniformly distributed in the search volume. The PDF for a single measurement z is then given by

$$p_z(z) = \frac{1}{V} \qquad 0 \le z \le V \tag{45}$$

where V is the measurement search volume in one dimension. Now consider the ML-PMHT LLR for a single, two-dimensional measurement (this would be time-delay/range and azimuth for active sonar). The LLR for this case can be written as

$$\Lambda_1(z_1, z_2) = \log\left(1 + \frac{\pi_1}{\pi_0}\frac{V}{\sqrt{|2\pi \mathbf{R}|}}\exp\left\{-\frac{1}{2}\left[\frac{(z_1-\mu_1)^2}{\sigma_1^2} + \frac{(z_2-\mu_2)^2}{\sigma_2^2}\right]\right\}\right) \tag{46}$$

Here $\mathbf{R}$ is the measurement covariance matrix, z_1 and z_2 are the measurements (time-delay and azimuthal), σ_1^2 and σ_2^2 are the time-delay and azimuthal measurement covariances, and finally, μ_1 and μ_2 are the (unknown) expected target locations in time-delay and bearing space.

The key here is to treat (46) as simply a random-variable transformation; i.e.,

$$w = \Lambda_1(z_1, z_2) \tag{47}$$

where z_1 and z_2 are uniform random variables, and w is a transformed random variable. This transformation can be done in several steps using the standard one-to-one random variable transformation technique: if X and Y are random variables, and $y = f(x)$, then

$$y = f^{-1}(y)\left|\frac{dx}{dy}\right| \tag{48}$$

This transformation is somewhat involved and must be done in stages; the step-by-step derivation is presented in [23]. The final result, where a single (2-dimensional) uniformly distributed measurement is transformed by the single-measurement ML-PMHT LLR (46), is given by

$$p_w(w) = \begin{cases} C\delta(w) & w = 0 \\ 2\pi \frac{\sigma_1\sigma_2}{V_1 V_2}\frac{\exp(w)}{\exp(w)-1} & 0 < w \le \log(1+K) \end{cases} \tag{49}$$

[2] This entire section is a summary of the work presented in [23].

where δ is a Dirac delta function, C is a normalization constant that is scaled so that the entire PDF integrates to 1, and

$$K = \frac{\pi_1}{\pi_0} \frac{V_1 V_2}{\sqrt{|2\pi \mathbf{R}|}} \tag{50}$$

The expression in (49) is the PDF of a single clutter measurement transformed by the ML-PMHT LLR; in this expression, the amplitude likelihood ratio $\rho_a(a)$ was assumed to be unity. It is possible to compute a closed-form expression in a similar fashion for a non-unity value of $\rho_a(a)$ for a fluctuating Gaussian model where the amplitude likelihood ratio is given by (39). The derivation is similar to that given above; the single-measurement transformed PDF (that includes the amplitude likelihood ratio) is

$$p_w(w) = \begin{cases} C\delta(w) & w = 0 \\ \frac{2\pi\sigma_1\sigma_2}{V_1 V_2} \left[1 - e^{\tau_\zeta} \left(\frac{e^w - 1}{K'}\right)^{\frac{1}{K_\sigma}}\right] \frac{\exp(w)}{\exp(w)-1} & \\ \qquad 0 < w \leq \log\left(1 + K' e^{-K_\sigma \tau_\zeta}\right) \end{cases} \tag{51}$$

Here

$$K' = \frac{K}{\sigma^2} e^{K_\sigma \tau_\zeta} \tag{52}$$

where σ^2 is the expected target intensity and K_σ is given by

$$K_\sigma = \frac{1 - \sigma^2}{\sigma^2} \tag{53}$$

Example plots of both (49) and (51) are shown in Fig. 1. It is instructive to look at the shape of these PDFs. The area under the blue curve in this figure can be roughly approximated as $2.5 \times 1e - 3 \approx 0.003$; the red curve, to first order, has about the same area. In other words, PDFs are roughly 99.7% delta functions located at $w = 0$. Because of this, even when adding many, many of these random variables together, the Central Limit Theorem will not apply.

3.2 *Batch Random-Variable Transformations*

At this point, the expression in (49) (or in (51)) gives the PDF for a single clutter measurement converted by the ML-PMHT LLR. As a tracker though, ML-PMHT is typically implemented as a batch tracker; it will process N_w scans, and each scan contains m_i measurements. For simplicity, let $M = \sum_{i=1}^{N_w} m_i$—here M is

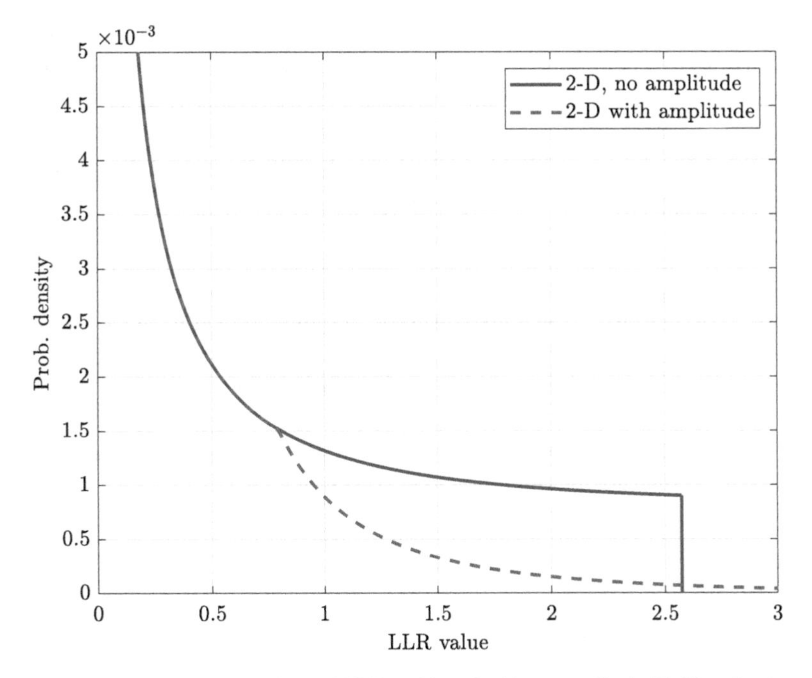

Fig. 1 Single measurement transformed PDFs, with and without amplitude likelihood ratio

the total number of measurements processed by the tracker. Since one of the ML-PMHT assumptions is that the measurements, conditioned on the target state, are independent, then the RVs that represent the single-measurement RV transformation of a clutter measurement will be independent as well. Thus, the PDF for a batch of measurements is simply computed by convolving (49) with itself $M - 1$ times. An alternative approach (one that is slightly more difficult but is computationally faster and more accurate for large values of M) is to take the characteristic function of the single measurement PDF, exponentiate it by M, and then take the inverse characteristic function.

3.3 Peak PDFs

The ML-PMHT batch PDF described in the previous section is used to compute the peak PDF—the distribution that describes the *maximum* point in the LLR that is caused by a batch of clutter measurements. To do this, extreme-value theory [24–29] is used. Work in [30] shows that the peak point in the ML-PMHT LLR is well described by a Gumbel distribution. This distribution takes the form of

$$f(x) = \frac{1}{\beta} \exp\left[-\left(\frac{x-\nu}{\beta}\right) - \exp\left(-\frac{x-\nu}{\beta}\right)\right] \tag{54}$$

In an actual ML-PMHT tracker implementation, a numerical global optimization scheme is used to determine the peak point in the ML-PMHT LLR. In order to determine the parameters in the Gumbel distribution given by (54), it is necessary to model the effect of the numerical optimization used in the actual tracker implementation. One conceptually simple way to find the maximum of the ML-PMHT LLR would be to sample the LLR with a very finely spaced grid, consisting of M_{tot} samples, and take the maximum from the sample set as the global maximum value of the LLR. This is obviously not done in practice because M_{tot} becomes prohibitively large, especially when the dimension of the state vector $\mathbf{x}$ is large. However, it is possible to make some basic assumptions about the LLR surface and determine a value for the number of samples M_{tot} that could be theoretically required to get the same accuracy on the global maximum as the actual numerical optimization scheme.

It turns out that a good approximation to this number is given for each measurement dimension by the following expression:

$$M_{tot} = \frac{1}{2}\sqrt{\frac{m_{ave}V}{\sigma\varepsilon}\frac{K}{K+1}} + 1 \tag{55}$$

where m_{ave} is the average number of clutter measurements in a scan, and ε is the assumed accuracy of the numerical optimization scheme.

With this "equivalent" value of M_{tot}, it is possible to compute the parameters β and ν in the Gumbel distribution given by (54). Now, if $F(x)$ is the cumulative distribution function (CDF) being sampled (i.e., the integral of the ML-PMHT batch PDF described in Sect. 3.2), then the parameters ν and β can be obtained with the following [24]:

$$\nu = F^{-1}\left(1 - \frac{1}{M_{tot}}\right) \tag{56}$$

and

$$\beta = F^{-1}\left(1 - \frac{1}{eM_{tot}}\right) - \nu \tag{57}$$

As the preceding equation indicates, in order to obtain the Gumbel distribution parameters ν and β, it is necessary to obtain the CDF for the ML-PMHT batch PDF. This CDF must be obtained and then inverted numerically since it has no convenient closed-from expression. For numerical reasons, the most accurate way to do this is to perform a cumulative numerical integration of the ML-PMHT batch PDF and then approximate the right-hand side of this with the following relatively simple closed-form expression:

$$F_{approx}(x) = 1 - \exp\left[-k(x-m)^{l}\right] \tag{58}$$

The parameters $k, m,$ and l are fit to the numerical CDF in the region $F(x) > 0.95$ using numerical optimization. With these values calculated, it is straightforward to invert (58). Letting $F_{approx}(x) = \phi$ and solving for x produces

$$x = m + \left[-\frac{1}{k} \log(1 - \phi) \right]^{\frac{1}{l}} \tag{59}$$

Now, combining (56), (57), and (59) gives a final expression for the Gumbel distribution parameters

$$\nu = m + \left[\frac{1}{k} \log(M_{tot}) \right]^{\frac{1}{l}} \tag{60}$$

and

$$\beta = \left[\frac{1}{k} + \frac{1}{k} \log(M_{tot}) \right]^{\frac{1}{l}} - \left[\frac{1}{k} \log(M_{tot}) \right]^{\frac{1}{l}} \tag{61}$$

With this, it is possible to write the Gumbel distribution that represents the peak in the ML-PMHT LLR surface caused by clutter. Figure 2 is an example of this—the green curve is a batch clutter PDF, and the magenta curve is the resultant peak clutter PDF. Also shown is the calculated tracking threshold that would produce a false track rate of $P_{FT} = 0.01$.

4 Miscellaneous

In this section, we touch on three miscellaneous subjects that will be useful in practical ML-PMHT implementations. They are (1) calculating the CRLB for ML-PMHT; (2) a multitarget ML-PMHT implementation; (3) a maneuver-model parameterization for both ML-PDA and ML-PMHT. These topics are only summarily discussed; for detailed information on implementation issues the interested reader should consult the provided references.

4.1 ML-PMHT Crámer-Rao Lower Bound

All tracking algorithms produce an estimate of the target state (this is, of course, the point of such algorithms); however, not all algorithms provide an estimate of the *accuracy* of the state estimate. In real-world applications, some idea of the accuracy of an estimate is of critical importance. For instance, the Kalman filter produces both a state estimate and a direct state covariance estimate. Unfortunately,

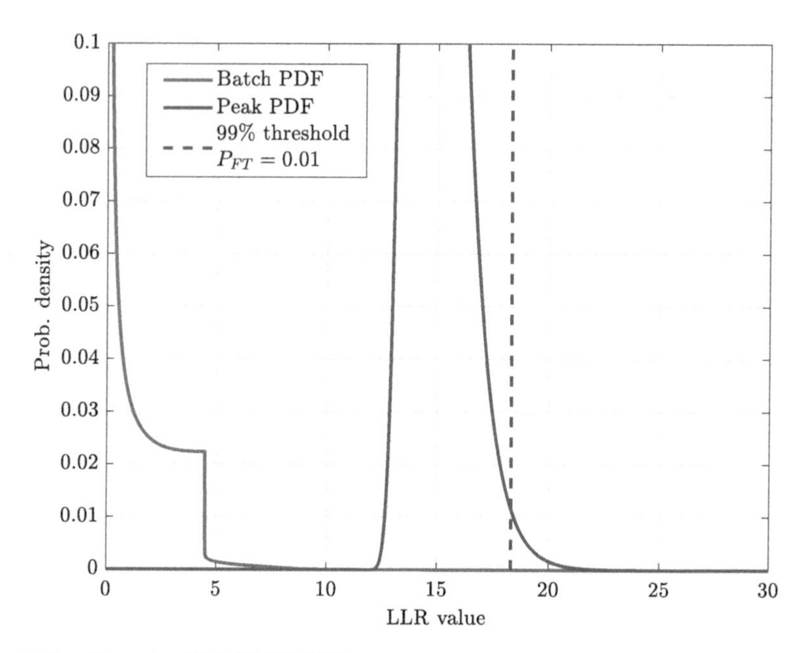

Fig. 2 Two-dimensional ML-PMHT PDFs

neither ML-PDA nor ML-PMHT provide a direct covariance estimate. However, both algorithms have been shown to be efficient estimators [31, 32]—that is, the accuracy of their estimates meets their CRLB. Thus, the CRLB matrix can be used as an estimate of the state covariance.

The form of the LLR for ML-PDA forces the computation of its CRLB to be a very high-dimensional integral [31]—on the order of the number of measurements in a scan. Thus, it needs to be computed off-line with Monte-Carlo integration, which is not practical for a real-time tracker implementation. Fortunately, the form of the LLR for ML-PMHT allows for many cancellations in the CRLB integral, which makes it tractable in real time. (This is another significant advantage for ML-PMHT over ML-PDA.)

The CRLB for the ML-PMHT can be calculated via the Fisher Information Matrix (FIM). The FIM for one scan of data is given by [32]

$$\mathbf{J}_i = \mathbf{D}_\phi^T \sum_{j=1}^{m_i} \mathbf{G}_j^T \int_V \frac{\frac{[\pi_1 p_t^\tau(a_j)]^2}{|2\pi \mathbf{R}_j|} e^{-\boldsymbol{\xi}_j^T \boldsymbol{\xi}_j} \boldsymbol{\xi}_j \boldsymbol{\xi}_j^T}{\frac{\pi_0 p_c^\tau(a_j)}{V} + \frac{\pi_1 p_t^\tau(a_j)}{\sqrt{|2\pi \mathbf{R}_j|}} e^{-\frac{1}{2}\boldsymbol{\xi}_j^T \boldsymbol{\xi}_j}} d\boldsymbol{\xi}_j da_j \frac{\mathbf{G}_j}{|\mathbf{G}_j|} \mathbf{D}_\phi \tag{62}$$

Here ϕ is the state-to-measurement conversion, and D_ϕ is the Jacobian of ϕ. The matrix $\mathbf{G}_j$ is the Cholesky decomposition of the inverse measurement covariance matrix $\mathbf{R}_j^{-1}$ such that $\mathbf{G}_j^T \mathbf{G} = \mathbf{R}_j^{-1}$. The vector $\boldsymbol{\xi}$ is $N \times 1$, where N is the dimensionality of the measurements, and the differential $d\boldsymbol{\xi}$ is also N-dimensional.

The FIM for an entire batch is just the sum of the FIM calculations for each scan:

$$\mathbf{J} = \sum_{i=1}^{N_w} \mathbf{J}_i \tag{63}$$

Finally, the state covariance is just approximated as the inverse of $\mathbf{J}$, the Fisher Information Matrix for the batch of data.

4.2 *Multitarget ML-PMHT*

It is very difficult to extend ML-PDA (or any other algorithm that assumes at most one measurement in a scan can originate from a target) to multiple targets. The ML-PDA LLR was developed for two targets in [33]; for any more targets than this, developing the LLR becomes almost a futile exercise in combinatorics. However, because the ML-PMHT algorithm allows for multiple measurements to originate from a given target in a scan, it is *exceedingly* easy to formulate a multitarget version for ML-PMHT [32]. For K targets with state vectors $\mathbf{x}_1, \ldots, \mathbf{x}_K$, the multitarget LLR is expressed as

$$\Lambda'(\mathbf{x}, Z) = \sum_{i=1}^{N_w} \sum_{j=1}^{m_i} \log \left\{ \pi_0 + V \sum_{k=1}^{K} \pi_k p[\mathbf{z}_j(i)|\mathbf{x}_k] \rho_{jk}(i) \right\} \tag{64}$$

where π_k is the probability that a given measurement is from the kth target and $\sum_{k=0}^{K} \pi_k = 1$. The ease of formulating the LLR in this fashion is another major advantages of ML-PMHT over ML-PDA.[3] Details of this multitarget implementation are discussed in [32].

4.3 *Maneuver-Model Parameterization*

We finally present a parameterization that can be applied to ML-PDA or ML-PMHT (or any deterministic batch tracker for that matter)—this parameterization allows for a rapidly maneuvering target. The "typical" target parameterization used by a batch tracker is a straight line. The length of the batch is a trade-off—it should be long enough so that there is enough "integration" to allow target measurements to

[3] A common misconception is that (64) allows a measurement to originate from multiple targets (i.e., a measurement can be shared between targets); this is not correct. Consider a single term of the sum—this represents the likelihood ratio for a single measurement. It can either originate from clutter *or* target 1 *or* target 2 ... *or* target K.

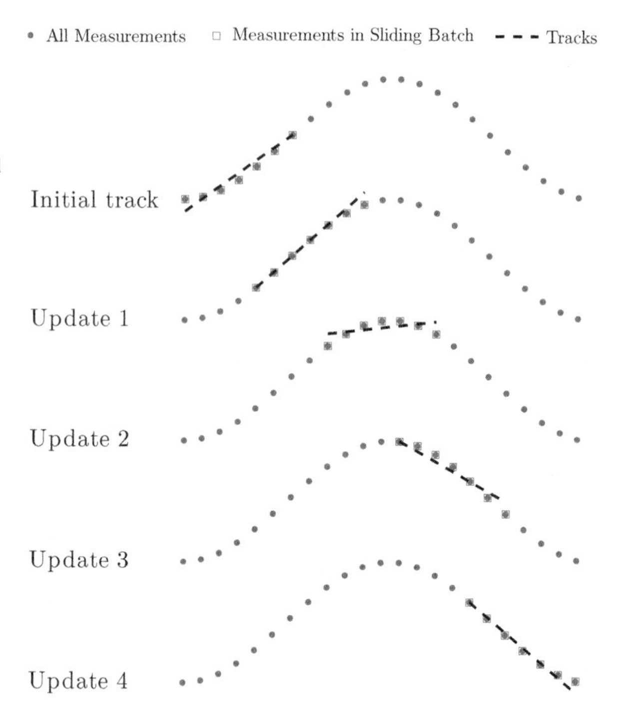

Fig. 3 Sliding window implementation for ML-PMHT in the case of a moderately maneuvering target. For the measurements in each sliding batch (denoted by blue squares), ML-PMHT straight-line solutions are shown. Reproduced with permission from [34], ©2013 IEEE

stand out from clutter measurements (i.e., long enough for the target measurements to form a recognizable line), but it should be short enough that any reasonable target path can be reasonably represented by a series of line segments. This is illustrated in Fig. 3. In this figure, the motion of a slowly maneuvering target can accurately be represented by a series of (overlapping) straight lines. The downfall to this approach occurs with rapidly maneuvering targets. If a target changes course very rapidly (i.e., quickly enough that the entire maneuver occurs between two measurements), then it is unrealistic to model the motion in a batch as a straight line. This situation is illustrated in Fig. 4. Here, the target maneuvers so rapidly that a batch tracker with a straight-line parameterization will invariably track off during the sudden target maneuvers. In order to enable ML-PMHT or ML-PDA to track targets performing these more severe maneuvers, we now add two components to the parameter vector, a maneuver time t_m and a maneuver angle (course change) θ_m. Now, the target motion in a batch can be represented as two line segments—this new parameterization is shown in comparison to the previous straight-line parameterization in Fig. 5. Details for exactly how this maneuver-model parameterization is implemented into ML-PMHT and ML-PDA are presented in detail in [34].

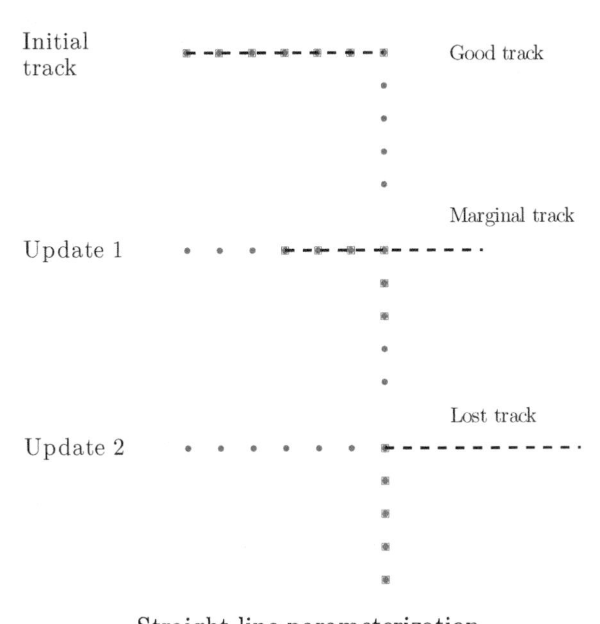

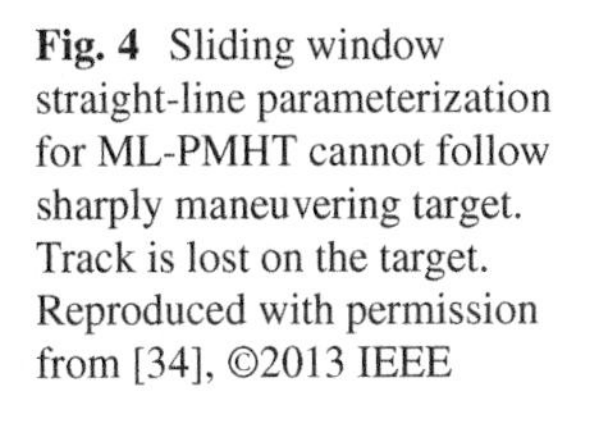
Fig. 4 Sliding window straight-line parameterization for ML-PMHT cannot follow sharply maneuvering target. Track is lost on the target. Reproduced with permission from [34], ©2013 IEEE

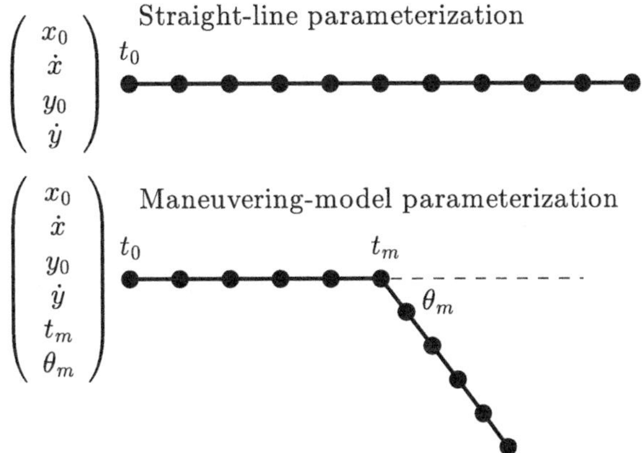

Fig. 5 Straight-line parameterization vs. maneuvering-model parameterization. Reproduced with permission from [34], ©2013 IEEE

5 Conclusion

In this chapter, we have presented two deterministic batch trackers—the Maximum Likelihood Probabilistic Data Association (ML-PDA) tracker and the Maximum Likelihood Probabilistic Multi-Hypothesis Tracker (ML-PMHT). The ML-PDA algorithm has been in the literature for a longer period of time, but recent work has shown that ML-PMHT has distinct advantages over ML-PDA.

The log-likelihood ratios for each algorithm, including two different amplitude likelihood models, were developed in detail. A method for determining tracking thresholds for ML-PMHT was then developed. Finally, a method for computing the CRLB for ML-PMHT was presented, a multitarget version of the ML-PMHT LLR was shown, and a maneuver-model parameterization for rapidly maneuvering targets was introduced. Together, this information should facilitate the implementation of the ML-PDA algorithm, and especially the ML-PMHT algorithm, or other deterministic batch trackers.

References

1. C. Jauffret, Y. Bar-Shalom, Track formation with bearing and frequency measurements in clutter. IEEE Trans. Aerosp. Electron. Syst. **26**(6), 999–1010 (1990)
2. M.R. Chummun, Y. Bar-Shalom, T. Kirubarajan, Adaptive early-detection ML-PDA estimator for LO targets with EO sensors. IEEE Trans. Aerosp. Electron. Syst. **38**(2), 694–707 (2002)
3. W. Blanding, P. Willett, S. Coraluppi, Sequential ML for multistatic sonar tracking, in *Proceedings Oceans 2007 Europe Conference*, Aberdeen (2007)
4. P. Willett, S. Coraluppi, Application of the MLPDA to bistatic sonar, in *Proceedings of the IEEE Aerospace Conference*, Big Sky (2005)
5. P. Willett, S. Coraluppi, MLPDA and MLPMHT applied to some MSTWG data, in *Proceedings of the 9th International Conference on Information Fusion*, Florence (2006)
6. Y. Bar-Shalom, P. Willett, X. Tian, *Tracking and Data Fusion: A Handbook of Algorithms*. YBS Publishing, Storrs (2011)
7. D. Avitzour, A maximum likelihood approach to data association. IEEE Trans. Aerosp. Electron. Syst. **28**(2), 560–566 (1996)
8. R. Streit, T. Luginbuhl, A probabilistic multi-hypothesis tracking algorithm without enumeration, in *Proceedings of the 6th Joint Data Fusion Symposium*, Laurel, MD (1993)
9. R. Streit, T. Luginbuhl, Maximum likelihood method for probabilistic multi-hypothesis tracking, in *Proceedings of the SPIE Conference on Signal and Data Processing of Small Targets, #2235*, Orlando, FL (1994)
10. R. Streit, T. Luginbuhl, Probabilistic multi-hypothesis tracking. Tech. Rep. TR 10428, Naval Undersea Warfare Center (1995)
11. P. Swerling, Probability of detection for fluctuating targets. IRE Trans. Inf. Theory **6**(2), 269–308 (1960)
12. J. Marcum, A statistical theory of target detection by pulsed radar. IRE Trans. Inf. Theory **6**(2), 59–267 (1960)
13. M. Richards, *Fundamentals of Radar Signal Processing* (McGraw-Hill, New York, 2005)
14. B. La Cour, Statistical characterization of active sonar reverberation using extreme value theory. IEEE J. Ocean. Eng. **29**(2), 310–316 (2004)
15. A. Jakeman, P. Pusey, A model of non-Rayleigh sea echo. IEEE Trans. Antennas Propag. **24**(6), 806–814 (1976)
16. A. Jakeman, On the statistics of k-distributed noise. J. Phys. A Math. Gen. **13**, 31–48 (1980)
17. D. Abraham, A. Lyons, Novel physical interpretations of K-distributed reverberation. IEEE J. Ocean. Eng. **27**(4), 800–813 (2002)
18. D. Abraham, A. Lyons, Simulation of non-Rayleigh reverberation and clutter. IEEE J. Ocean. Eng. **29**(2), 347–362 (2004)
19. D. Abraham, Detection-threshold approximation for non-Gaussian backgrounds. IEEE J. Ocean. Eng. **35**(2), 355–365 (2010)

20. D. Abraham, J. Gelb, A. Oldag, Background and clutter mixture distributions for active sonar statistics. IEEE J. Ocean. Eng. **36**(2), 231–247 (2011)
21. K. Ward, Compound representation of high resolution sea clutter. Electron. Lett. **17**, 561–565 (1981)
22. A. Abramowitz, I. Stegun, *Handbook of Mathematical Functions*. National Bureau of Standards: Applied Mathematics Series, vol. 55, Chaps. 9.1.1, 9.1.98, and 9.12 (Dover, New York, 1965)
23. S. Schoenecker, P. Willett, Y. Bar-Shalom, Extreme-value analysis for ML-PMHT, part 1: threshold determination for false track probability. IEEE Trans. Aerosp. Electron. Syst. **50**(4), 2500–2514 (2014)
24. E. Castillo, *Extreme Value Theory in Engineering* (Harcourt Brace Jovanovich, Boston, 1988)
25. S. Coles, *An Introduction to Statistical Modeling of Extreme Values* (Springer, London, 2001)
26. H. David, *Order Statistics* (Wiley, New York, 1981)
27. P. Embrechts, C. Klüppelberg, T. Mikosch, *Modelling Extremal Events for Insurance and Finance* (Springer, Berlin, 1997)
28. J. Galambos, *The Asymptotic Theory of Extreme Order Statistics* (Wiley, New York, 1978)
29. E. Gumbel, *Statistics of Extremes* (Columbia University Press, New York, 1958)
30. W. Blanding, P. Willett, Y. Bar-Shalom, Offline and real-time methods for ML-PDA track validation. IEEE Trans. Signal Process. **55**(5), 1994–2006 (2007)
31. C. Jauffret, Y. Bar-Shalom, Track formation with bearing and frequency measurements in clutter, in *Proceedings of the 29th Conference on Decision and Control*, Honolulu (1990)
32. S. Schoenecker, P. Willett, Y. Bar-Shalom, ML-PDA and ML-PMHT: comparing multistatic sonar trackers for VLO targets using a new multitarget implementation. J. Ocean. Eng. **39**(2), 303–317 (2014)
33. W. Blanding, P. Willett, Y. Bar-Shalom, ML-PDA: advances and a new multitarget approach. EURASIP J. Adv. Signal Process. **2008**, 1–13 (2008)
34. S. Schoenecker, P. Willett, Y. Bar-Shalom, The ML-PMHT multistatic tracker for sharply maneuvering targets. IEEE Trans. Aerosp. Electron. Syst. **49**(4), 2235–2249 (2013)

Moving Horizons Estimation for Wheelchair Trajectory Repeatability in the Home

Steven B. Skaar

1 Introduction

A portion of the wheel-chair-bound population experiences difficulty navigating power wheelchairs through the home [1]. One move toward an autonomous chair, suitable for in-home transportation from an approximately known starting pose to a preselected new pose, began in the 1990s [2]. Using an adaptation of the "teach-repeat" paradigm that was used with robotic arms in factories, the attempt was made to automatically replicate in the home human-"taught" trajectories.

The idea as used in factories—for spot welding of auto frames, for example—was that "workpieces," the objects of the robot's operation, such as car-frame members to be joined, could be positioned or "fixtured," copy after copy, in the same three-dimensional locations in order to allow a robot to succeed in its purpose by undergoing repetition of previously "taught" trajectories of the robotic arm's tool-bearing "end effector" [3]. The application of a similar approach to wheelchair control is simplified—relative to the spot-welding application, for example—from the fact that, typically, the taught and repeated trajectories would be two- rather than three-dimensional, that is, the chair is constrained to remain on a relatively flat floor. Also, the wheelchair application would not require the complementary, often-costly active pre-fixturing of workpieces [4].

The problem introduced by the wheelchair application, however, lies with the nonholonomic relationship [5] between rotation of the mobile robot's directly measurable/controllable wheels and the position/orientation of the transported body.

S. B. Skaar (✉)
Department of Aerospace and Mechanical Engineering, University of Notre Dame, Notre Dame, IN, USA
e-mail: skaar.1@nd.edu

A. A. Ruffa, B. Toni (eds.), *Advanced Research in Naval Engineering*, STEAM-H: Science, Technology, Engineering, Agriculture, Mathematics & Health, https://doi.org/10.1007/978-3-319-95117-1_6

This means that return of the wheels' angles of rotation to that series associated with the taught trajectory is, unlike the spot welding example of an industrial, holonomic robot, no guarantee of return of the chair to its associated, desired sequence of poses in the plane of the floor. Because the relationship between these real-time-measurable internal angles and the pose of the chair—even in the absence of wheel slip—is modeled at best differentially, the kind of control used in industry of the robotic arm's joint rotations is inadequate for controlling wheel rotation. Lack of a priori integrability of the Pfaffian form [5] of the kinematic constraints means there is a path-dependent/tracking-dependent component of the outcome that has no counterpart in the industrial-robot case. With the further reality of wheel slip, the proposition of teach/repeat becomes even more problematic because even numerical integrals based on measured wheel rotations and the differential kinematics generally lose accuracy with distance traveled.

The Extended Kalman Filter (EKF) [6], an algorithm that has been used to apply "observations," in order, for example, to correct and improve state estimates of "inertial navigation" (which estimates are based on measured acceleration), was initially employed for the wheelchair [2]. Just as inertial navigation uses acceleration measurements that are related to time derivatives of 3D aircraft position and orientation in space, so too the wheelchair's measured wheel rotations are related differentially to its evolving 2D position and orientation in the plane of the floor. Both applications require, in order to avoid degradation of estimates with time or distance traveled, that "observations," which are related to the instantaneous state via "observation equations," alter the numerical integrals based upon differences between expected and actual observations. State equations and observation equations in both the aerospace and wheelchair applications are generally nonlinear.

Observations used in the EKF are presumed to have random zero-mean error; they are acquired over time, in order, optimally (in the linear limit), to improve state estimates. This recursive "innovation" process offers the computational advantage of not requiring retention of a batch of past observations. Past information is instead retained approximately in the updated parameters of a Gaussian error-covariance matrix of the estimated state. Some researchers, however, have found that, with large nonlinearity in the state- and/or observation-equations, the alternative Moving Horizons Estimator (MHE) [7] improves upon the EKF's results, albeit at a computational cost and the burden of retaining a batch of data within a moving window of time (hence "Moving Horizons") of the estimated state. The improvement is due in part to imperfect data retention for the present, nonlinear case in the EKF's covariance matrix, or due to certain kinds of model error.

The present application, with its emphasis on "repeatability" rather than "accuracy," adds another reason to apply the MHE; it has to do with application locally of visual observations of physical fiducials (or "beacons" or "cues") placed during system set-up, as discussed in [2], on the walls of the home of the disabled user. Accumulation of a local batch of observations that is distributed along the route according to criteria developed herein can be enforced by reducing the chair's speed, if necessary, as required in the current tracking event. This accumulation should be distributed in accordance with the local region of the reference path as it was using a teaching event from which a similar batch is acquired.

Just as with some holonomic, industrial-robot applications, “repeatability” of taught poses can be sufficient for the task at hand (even if the more costly objective, “accuracy,” is not good), so too, here, we seek high-precision trajectory repeatability. The paper shows how MHE can bring this about even in a domestic environment where there is considerable deviation from a nominal geometric model. The present article compares issues of repeatability in the present, nonholonomic case with the better-known literature pertaining to repeatability of holonomic, industrial robots. The experimental run used to illustrate the system estimation and control can be seen on-line: https://youtu.be/7Yrc3IuXBus.

2 Process Equations

Unlike accelerometer-driven state equations, it is not necessary to use time (t) as the independent variable. In keeping with [2] we apply the kinematic variable α given by

$$\alpha = \frac{\theta_1 + \theta_2}{2} \tag{1}$$

where θ_1 and θ_2 are the forward wheel rotations of the two motor-driven wheels, in radians, as measured using optical encoders, per Fig. 1.

If θ_1 and θ_2 are both positive through the path leading to the ith observation in the short trajectory shown in Fig. 1, then the path length, s_i, shown in the figure, is, according to the nominal forward kinematics, given by

$$s = R\alpha \tag{2}$$

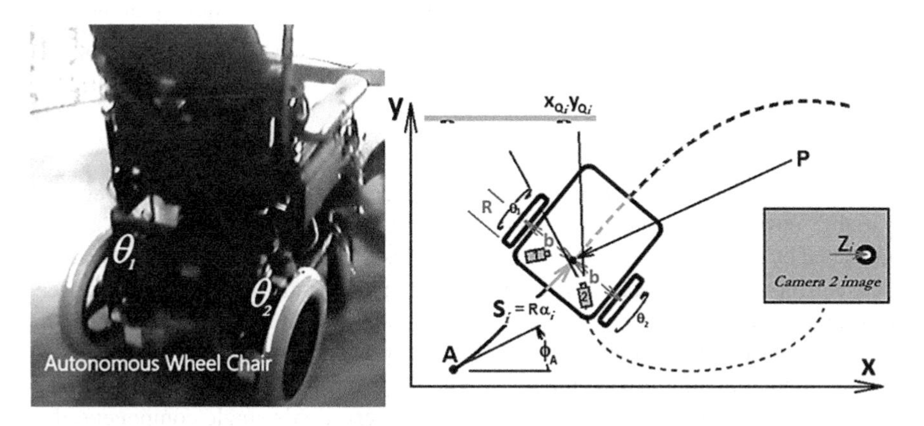

Fig. 1 Definition of angle-rotation coordinates; Camera 2 captures ith observation z_i

where, as indicated in Fig. 1, R is the radius of both of the two drive wheels. The choice of independent variable, $\alpha = (\theta_1 + \theta_2)/2$, was helpful in the EKF application because of the process noise [6] of the differential-equation plant model: EKF adds to each of the three kinematic state equations given below zero-mean white noise w:

$$\frac{dx_P}{d\alpha} = R\cos\phi_P + w_x(\alpha) \tag{3}$$

$$\frac{dy_P}{d\alpha} = R\sin\phi_P + w_y(\alpha) \tag{4}$$

$$\frac{d\phi_P}{d\alpha} = \frac{R}{b}u(\alpha) + w_\phi(\alpha) \tag{5}$$

where two of the three states, x_P, y_P, are the in-plane coordinates of point P on the chair, which is the midpoint between the two wheels as shown in Fig. 1, and where ϕ_P is the in-plane orientation of the chair's forward direction relative to the fixed x-y frame. The initial condition "ϕ_A" shown in Fig. 1 is the initial (starting point A) wheelchair orientation for the trajectory indicated. The "control variable" u is an indication of steering at any path juncture; in terms of the two measurable wheel rotations, θ_1 and θ_2:

$$u = \frac{\Delta\theta_2 - \Delta\theta_1}{\Delta\theta_2 + \Delta\theta_1} \tag{6}$$

in the infinitesimal limit where each incremental Δ becomes the differential d. (In practice, u is evaluated using the small but finite $\Delta\theta_1$ and $\Delta\theta_2$ as measured simultaneously by each wheel's optical encoder.) The introduced control variable u in the stochastic Eqs. (3)–(5) is related to the signed radius of curvature ρ of the path at location s according to

$$u = \frac{R}{\rho} \tag{7}$$

The choice of the kinematic variable α rather than time t as the independent variable ensures that, when the chair is at rest, EKF's estimate uncertainty as indicated by the state error-covariance matrix ceases to grow as time passes. Instead, between observations, state covariances change only with advances in distance traveled, $s = R\alpha$.

At the instant pictured in Fig. 1, one of several wall cues (see Fig. 2) is detected by the onboard, calibrated "Camera 2." The cue's nominal coordinates relative to the indicated reference frame are known. The observation's single component, the horizontal location of the cue in the indicated image, is denoted by z_i. This is the ith of a selected batch of n observations—z_1, z_2, ... z_i ... z_n—to be used in a form

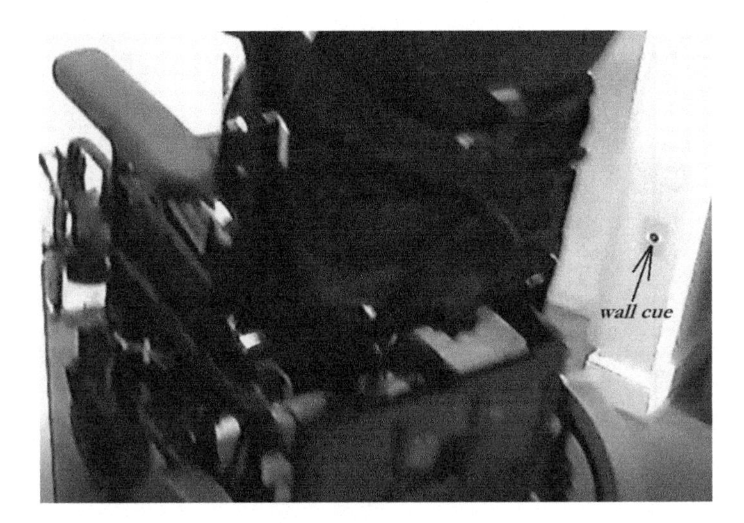

Fig. 2 Camera beneath seat acquires wall-cue image

of Moving Horizons Estimator (MHE). For purposes of the present discussion no generality is lost in considering the use of the n observations to generate the estimate of the *initial* state, x_A, y_A, ϕ_A, that minimizes:

$$\mathbf{J} = \mathbf{J}\left(x_A, y_A, \phi_A\right) = \frac{\mathbf{1}}{\mathbf{2}}\mathbf{r}^{\mathbf{T}}\mathbf{W}\mathbf{r} \tag{8}$$

where **W** is a diagonally dominant, symmetric, positive-definite, nxn weighting matrix and where the ith of n elements of the vector of residuals **r** is given by:

$$r_i = z_i - h_i\left(x_{Pi}\left(x_A, y_A, \phi_A\right), y_{Pi}\left(x_A, y_A, \phi_A\right), \phi_{Pi}\left(x_A, y_A, \phi_A\right); x_{Qi}, y_{Qi}\right) \tag{9}$$

where the nonlinear functional form h_i depends upon calibration parameters of the onboard camera used for the ith observation (Camera 1 or Camera 2 in Fig. 1) as well as the previously measured nominal in-plane position of the observed cue. The two parameterizing values, x_{Qi}, y_{Qi}, are these measured coordinates in the x-y plane of the floor of the particular fiducial or "cue" sampled for the ith observation. The focal axes of the two cameras are parallel to the floor and mounted such that they are at the same elevation above the floor as the center of the cues.

The dependency of the elements $x_{Pi}, y_{Pi}, \phi_{Pi}$ on initial conditions x_A, y_A, ϕ_A is nonlinear and requires numerical integration of the modeled nonholonomic kinematics, that is, the deterministic part of Eqs. (3)–(5). This includes the domain $0 < \alpha < s_i/R$ over which the batch of observations, including observation i, as well as starting point A is acquired.

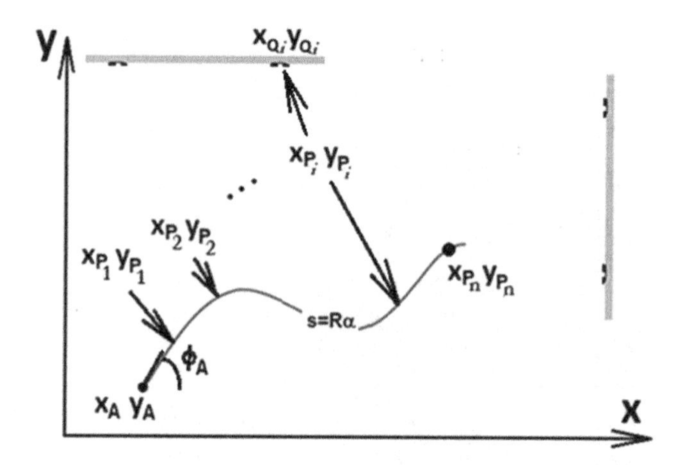

Fig. 3 Series of the batch of n observations

The nonlinearity of this part of the observation equation requires iteration, as indicated below, in order to minimize **J** with respect to initial conditions x_A, y_A, ϕ_A. This would ordinarily necessitate separate numerical integration for each iteration as estimated initial conditions converge sequentially with increments Δx_A, Δy_A, $\Delta \phi_A$ recalculated for each iteration. (A three-dimensional rigid-body counterpart to our problem, such as with satellite-attitude estimation [8], the inspiration for [2], would, for batch estimation, require repeated, iterative numerical integration of the state equations together with a 3x3 state-transition matrix $\Phi(t_i,t_A)$ from which the counterparts to Δx_A, Δy_A, $\Delta \phi_A$ are iteratively determined.)

Because of the integrals of Eqs. (3)–(5) developed below, however, the initial numerical integration is re-used as the batch algorithm undergoes iterative convergence. This (together with affordable, very fast, computation hardware) makes the MHE—computationally costly relative to EKF—practicable for the present real-time application. It is worth noting that the recursive (not batch) EKF *also* requires real-time numerical integration of the state equations. EKF also requires real-time numerical integration of covariance estimates of a 3x3 error-covariance matrix [6], which the present estimator does not. So, although EKF does not require retention of the typical batch size, n, of twenty or so observations, it does require other numerical/memory overhead not needed here.

Using the third of Eq. (1), ϕ_P can be integrated forward from ϕ_A through all of the n points where observations are acquired, as seen in Fig. 3. Thus, from Eqs. (5) and (6):

$$\phi_P(\alpha) = \phi_A + \int_0^{\alpha} \frac{R}{b} u(\lambda)\, d\lambda \approx \phi_A + \sum_{j=1}^{m} \frac{R\left(\Delta\theta_{2j} - \Delta\theta_{1j}\right)}{2b} \equiv \phi_A + I(\alpha) \tag{10}$$

where $\Delta\theta_{1j}$ and $\Delta\theta_{2j}$ are the jth of m wheel rotation increments, in radians, of wheels 1 and 2, simultaneously acquired. These come in typically very small encoder-measured increments, under 1/100th of a radian, so m is a large integer (much larger than the number n of camera/cue samples across the same horizon.)

The integral $I(\alpha)$ can be re-used in the iterative process to solve for the initial conditions, x_A, y_A, ϕ_A, that minimize $\mathbf{J}(x_A, y_A, \phi_A)$. The first correction, using Gauss–Newton iteration, is determined from $\mathbf{x}_\mathbf{A}^{(0)}$, the 3-element initial guess for x_A, y_A, ϕ_A, according to [8]

$$\mathbf{x}_\mathbf{A}^{(1)} = \mathbf{x}_\mathbf{A}^{(0)} + [\boldsymbol{A}^T\boldsymbol{W}\boldsymbol{A}]^{-1}\boldsymbol{A}^T\boldsymbol{W}\mathbf{r} \tag{11}$$

where $\mathbf{r}$ and $\boldsymbol{W}$ are, respectively, n sample residuals and the nxn weighting matrix, mentioned above; and where $\boldsymbol{A}$ is a 3-column, n-row matrix whose ith-row elements are given by

$$A_{i1} = \frac{\partial hi}{\partial \boldsymbol{x_A}} \mid \mathbf{x}_\mathbf{A}^{(0)} \tag{12}$$

$$A_{i2} = \frac{\partial hi}{\partial \boldsymbol{y_A}} \mid \mathbf{x}_\mathbf{A}^{(0)} \tag{13}$$

$$A_{i3} = \frac{\partial hi}{\partial \boldsymbol{\phi_A}} \mid \mathbf{x}_\mathbf{A}^{(0)} \tag{14}$$

Evaluation of these partial derivatives for the ith observation equation, *hi*, entails the estimated position and orientation of the camera of the observation, and the sensitivity of that estimated camera position/orientation to increments in the initial-condition estimates given by Δx_A, Δy_A, $\Delta\phi_A$.

Consider the initial angular increment $\Delta\phi_A$, shown in Fig. 4a, associated with the batch of samples indicated in Fig. 3. The corresponding planar, rigid-body rotation of the odometry-based path allows for efficient evaluation of the partial derivative of Eq. (14). In particular, the increment in the estimated angular position of the vehicle at the juncture of the ith observation is the same everywhere across the interval of the batch, while the increment in the vehicle wheel-midpoint position is a simple, planar, rigid-body rotation of the entire path, such that, in the limit as $\Delta\phi_A$ approaches zero:

$$\Delta x_{Pi} = -y_{rel}\ \Delta\phi_A \tag{15}$$

$$\Delta y_{Pi} = x_{rel}\ \Delta\phi_A \tag{16}$$

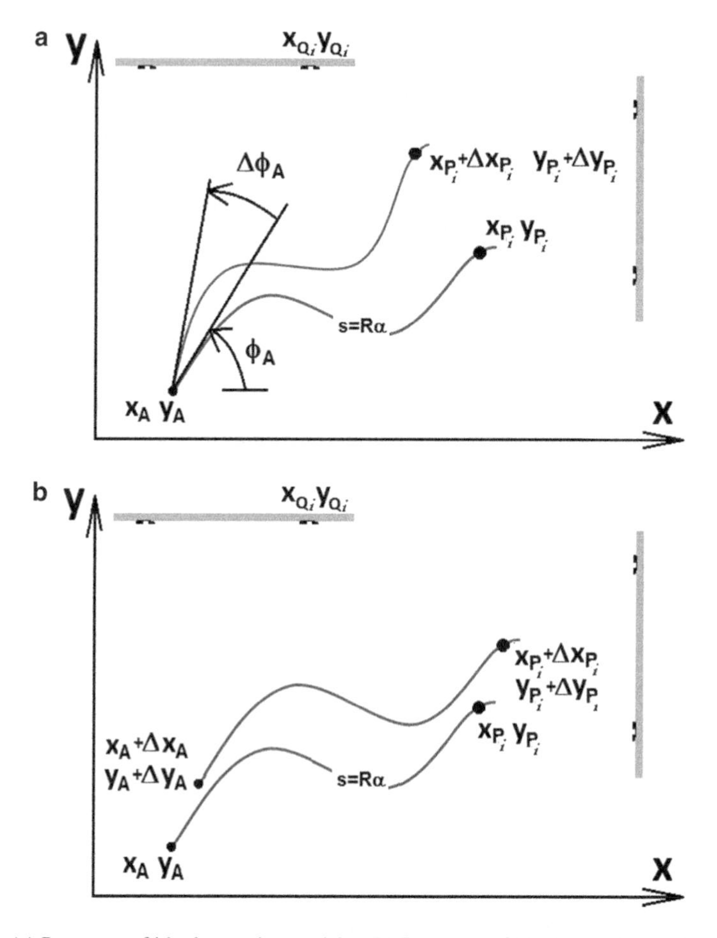

Fig. 4 (**a**) Response of ith observation to $\boldsymbol{\Delta\phi_A}$. (**b**) Response of ith observation to $\boldsymbol{\Delta x_A, \Delta y_A}$

where $x_{rel} = (x_{Pi} - x_A)$ and $y_{rel} = (y_{Pi} - y_A)$ for P_i, the position of the midpoint between the wheels, at the moment of the ith observation within the batch.

Consider a corresponding increment, Δx_A, Δy_A, in the estimated position of point A, as shown in Fig. 4b. If the entire path moves as a rigid body with this increment, then the corresponding change in the estimate of the position of the midpoint between the wheels at juncture i due to this move is

$$\Delta x_{Pi} = \Delta x_A \tag{17}$$

$$\Delta y_{Pi} = \Delta y_A \tag{18}$$

Because each of the two cameras depicted in Fig. 1 move as a rigid body with the vehicle, Eqs. (17) and (18) also describe sensitivity in the plane of the floor of estimates of each camera's focal point.

With known coordinates x_{Qi}, y_{Qi} of the cue detected in the ith observation, the observation equation h_i of Eq. (9) can be written, for the simple pinhole camera model [2] used herein:

$$h_i\left(x_{Pi}, y_{Pi}, \phi_{Pi}; x_{Qi}, y_{Qi}\right) = \mathrm{f}\,\mathrm{N}\left(x_{Pi}, y_{Pi}, \phi_{Pi}; x_{Qi}, y_{Qi}\right)/\mathrm{D}\left(x_{Pi}, y_{Pi}, \phi_{Pi}; x_{Qi}, y_{Qi}\right) \quad (19)$$

where f is the single intrinsic camera-calibration parameter, and where functions N and D are parametrized by the aforementioned extrinsic camera-calibration parameters. In light of Eqs. (17) and (18) the derivatives of Eqs. (12) and (13) can be written:

$$A_{i1} = \frac{\partial hi}{\partial x_A}\Big|_{\mathbf{x}_A^{(0)}} = \mathrm{f}\,(\mathrm{N}_{x_{Pi}}(x_{Pi}, y_{Pi}, \phi_{Pi}; x_{Qi}, y_{Qi})/\mathrm{D} - \mathrm{N}\,\mathrm{D}_{x_{Pi}}(x_{Pi}, y_{Pi}, \phi_{Pi}; x_{Qi}, y_{Qi})/\mathrm{D}^2)\Big|_{\mathbf{x}_A^{(0)}} \quad (20)$$

$$A_{i2} = \frac{\partial hi}{\partial y_A}\Big|_{\mathbf{x}_A^{(0)}} = \mathrm{f}\,(\mathrm{N}_{y_{Pi}}(x_{Pi}, y_{Pi}, \phi_{Pi}; x_{Qi}, y_{Qi})/\mathrm{D} - \mathrm{N}\,\mathrm{D}_{y_{Pi}}(x_{Pi}, y_{Pi}, \phi_{Pi}; x_{Qi}, y_{Qi})/\mathrm{D}^2)\Big|_{\mathbf{x}_A^{(0)}} \quad (21)$$

where the subscripts of N and D denote partial derivatives of these length expressions with respect to x_{Pi} or y_{Pi}.

Likewise, the estimate of the angular position of each camera increments, along with the angular increment of the entire wheelchair base, as $\Delta\phi_A$. Evaluation of A_{i3} entails the response of the sample z_i both to orientation change of the camera consistent with $\Delta\phi_A$ and to position change of the camera's focal point as a consequence of $\Delta\phi_A$, per Eqs. (15) and (16) above.

$$A_{i3} = \frac{\partial hi}{\partial \phi_A}\Big|_{\mathbf{x}_A^{(0)}} = \mathrm{f}\,(\mathrm{N}_{\phi_{Pi}}(x_{Pi}, y_{Pi}, \phi_{Pi})/\mathrm{D} - \mathrm{N}\,\mathrm{D}_{\phi_{Pi}}(x_{Pi}, y_{Pi}, \phi_{Pi})/\mathrm{D}^2)\Big|_{\mathbf{x}_A^{(0)}} + y_{rel}\,A_{i1} - x_{rel}\,A_{i2} \quad (22)$$

where y_{rel} x_{rel}denote the estimated position of the camera's focal point with respect to the estimate of point A, referred to the room's fixed y and x reference frames, respectively.

Subsequent iterations through to convergence are computed quickly

$$\mathbf{x}_A^{(k)} = \mathbf{x}_A^{(k-1)} + [\boldsymbol{A}^T\boldsymbol{W}\boldsymbol{A}]^{-1}\boldsymbol{A}^T\boldsymbol{W}\mathbf{r} \quad (23)$$

using the same once-integrated path of Figs. 3, 4, and 5 and successively defined changes in the in-plane coordinates and orientation of the initial condition at A. Each

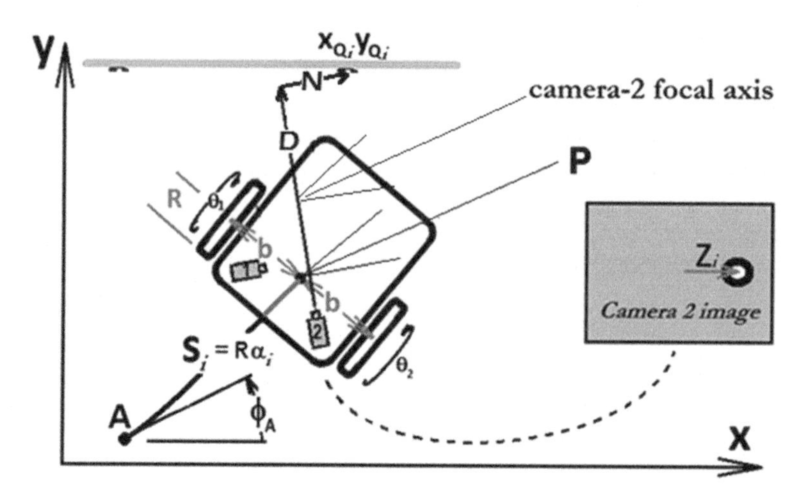

Fig. 5 Estimate of Camera 2's focal-point position in the plane of the floor, using previously found extrinsic calibration parameters, changes together with estimated initial-condition changes, $\mathbf{\Delta x_A}$, $\mathbf{\Delta y_A}$

repetition of Eq. (23) also requires recalculation, for each sample i in the batch, of the vehicle coordinates at the wall-cue-sample path junctures:

$$new\ \phi_{Pi} = old\ \phi_{Pi} + \Delta\phi_A \tag{24}$$

$$new\ x_{Pi} = old\ x_{Pi} + \Delta x_A - \sin\left(\Delta\phi_A\right) * y_{rel} - \left(1 - \cos\left(\Delta\phi_A\right)\right) * x_{rel} \tag{25}$$

$$new\ y_{Pi} = old\ y_{Pi} + \Delta y_A + \sin\left(\Delta\phi_A\right) * x_{rel} - \left(1 - \cos\left(\Delta\phi_A\right)\right) * y_{rel} \tag{26}$$

where x_{rel} and y_{rel} locate P_i relative to A, according to the outgoing estimates. The functions $\sin(\Delta\phi_A)$ and $\cos(\Delta\phi_A)$ are computed once for all n observations, helping make this calculation compatible with real time.

Point A though identified above with an "initial condition" could be defined as any point along a path segment that includes the batch of samples; for purposes of real-time trajectory tracking, it is considered to be the instantaneous location of P on the vehicle at the moment when the iterative batch calculation is commenced. Ongoing wheel-rotation samples are used to transition the converged estimate to the current point; a PID control algorithm steers the vehicle toward a corresponding reference-path juncture. The reference path, in turn, is determined as discussed below based on the result of a similar exercise conducted over the course of a prior teaching event, where a human pushes the chair through the path which is to be repeated.

Of primary practical interest is system repeatability: How closely and reliably does the chair come to repeating the trajectory as taught? The remainder of the present work is directed toward evaluation of the relationship between repeatability and the choice and acquisition during tracking of the batch of samples used to create estimates at various junctures throughout the trajectory of interest.

3 Repeatability

For holonomic, industrial, robotic arms, repeatability is often considered the most critical system attribute; it is sometimes achieved everywhere across the extent of the end effector and tool to within a few hundredths of a millimeter. For the wheelchair, repeatability of the tracked path is similarly important since clearance of the outermost portions of the vehicle/rider combination, with respect to stationary walls and fixtures, can be tight, requiring, for example, repetition of the taught path to within a few millimeters across the extent of the moving wheelchair.

Repeatability for the vehicle depends largely on the estimation process. Selection of the batch of samples can be achieved in a way that promotes reliable repeatability precision approaching the level of millimeters according to the analysis below.

Figure 6 is a graphical representation of the reference path of https://youtu.be/7Yrc3IuXBus. This trajectory is used to illustrate the repeatability issues. Using the same batch approach during teaching as is discussed below in the context of tracking, best estimates of x_P, y_P, and ϕ are determined at each nodal point indicated in Fig. 6. Each interval, or leg, between consecutive nodal points is approximated

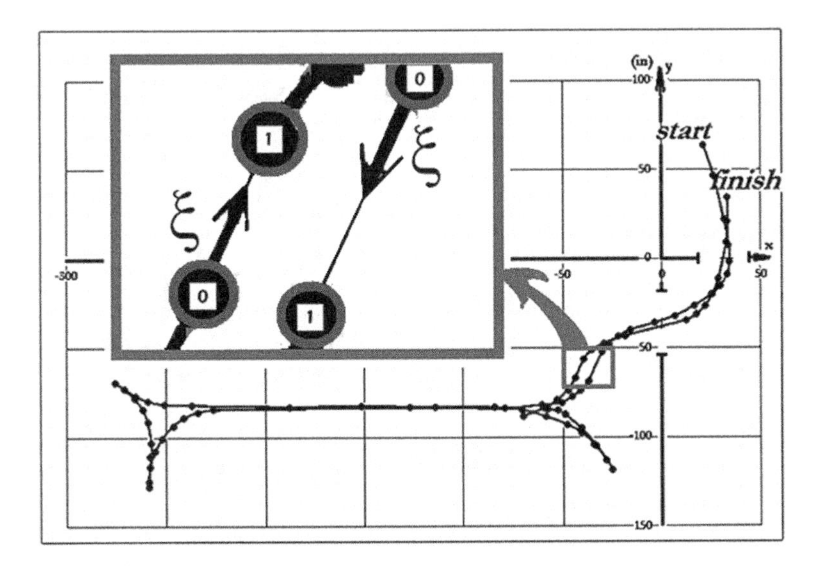

Fig. 6 Reference path used in http://youtu.be/7Yrc3luXBus

with a straight line. A dimensionless coordinate ξ is defined to range from 0 to 1, as the reference moves across the leg in the direction of the wheelchair movement. The reference path at intermediate leg junctures interpolates linearly between nodes, with ξ.

Because time t is not used as the independent variable, the reference path advances by sequential leg advancement and by a changing ξ within a leg. At any juncture, the current estimate of x_P, y_P, and ϕ is used to find ξ within the current leg by identifying the intersection between the straight line of the leg and the estimated wheelchair-axle line that passes through the current estimate of x_P, y_P. Whenever ξ calculated in this way becomes greater than 1 the reference-path leg is advanced.

Target, or "reference" speed, $R(d\theta_1/dt + d\theta_2/dt)/2$, during tracking is set separately, based in part on reference-path curvature. The steering, or "control," $u = (d\theta_2/dt - d\theta_1/dt)/(d\theta_1/dt + d\theta_2/dt)$, is calculated based on a PID-like controller, using both the reference path and the estimated, current x_P–y_P–ϕ sequence. Together, these relations are used to calculate inputs to the two separate driving motors [2].

Repeatability depends in part on the effectuality of this control scheme—in particular the ability to track accurately the taught reference, or target, path. Herein, however, we focus on a different aspect of repeatability, one that pertains to estimation. In particular, one necessary condition for accurate repeatability can be tested directly using historical runs such as that of https://youtu.be/7Yrc3IuXBus.

The necessary condition is as follows: Suppose that estimates of an instantaneous x_P, y_P, and ϕ within a given tracking event are based upon a particular batch of n observations—$z_1, z_2, \ldots z_i \ldots z_n$—acquired shortly before that point in accordance with a proposed scheme for identifying the batch. Suppose further that an additional, completely different batch of observations—$z'_1, z'_2, \ldots z'_i \ldots z'_n$—from the same event, likewise selected in accordance with the proposed moving-horizons sample-selection scheme, is separately used to estimate x_P, y_P, and ϕ at the same juncture. To within any specified tracking accuracy, the two estimates must be the same.

4 Observability

The ability of the algorithm to satisfy the necessary condition of generating the same estimates with two distinct, qualifying batches of samples depends in part on the "observability" question [6]. For example, even in the case of perfect kinematic and observation modeling of the system, if even a very large number of elements of $\underline{z}$ and $\underline{z}'$ entail observation of just one particular cue, then the state is not observable. (An infinite number of paths, encircling the observed cue, could account identically well for the same observations.)

If Cues A through G of Fig. 7 were located at distances from the path far greater than shown, observability of the state would suffer. Even if these many, well-separated cues were to be detected with the same density per distance traveled as

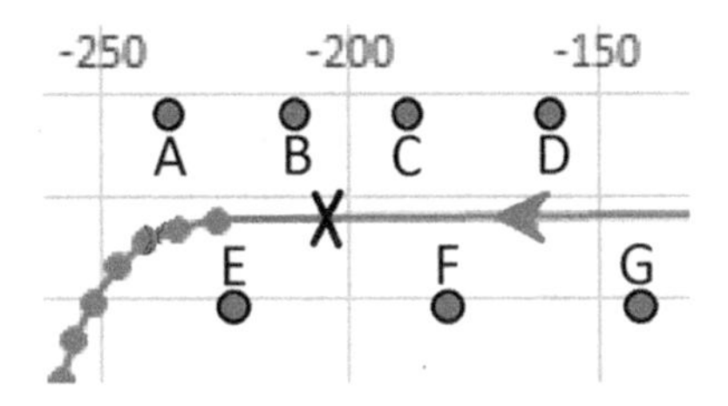

Fig. 7 First test portion of trajectory

per the current maneuver, observability would approach zero in the limit as distances approach infinity. Thus, other things being equal, observability is better if/as cues are close to the passing vehicle.

Additionally, if just two cues are represented in the batch, but there is no movement of the chair between samples, then, even though they may be located close to the vehicle, the state is not observable. If two cues are detected and there is some, but relatively little, movement of the base, then observability is poor. Consider, for example, the trajectory segment shown in Fig. 7, and pose estimates determined during a tracking event at the juncture denoted by X on the path. For purposes of illustration, we restrict the samples applied to the estimates x_P, y_P, ϕ to the final four, denoted here by—z_1 z_2 z_3 z_4—and entailing Cues C, E, and F only, as indicated in the top half of Fig. 8. Each of the five estimates entails a different group of samples of these three cues acquired during the same run. The five are contained within a 0.1-inch-radius circle.

In the bottom half of the same figure are five estimates, also acquired during this run, entailing just two of the three cues: C and E. Reduced observability is implied by the larger, 0.4-inch region across which the four estimates are strewn.

The comparison of the top versus the bottom halves of Fig. 8 is predicted by the 3x3 symmetric matrix $\boldsymbol{A^TWA}$, of Eq. (11), as follows: The eigenvalues of this matrix, $\lambda_1 > \lambda_2 > \lambda_3$, are positive and real, and relate to the condition number [9] according to

$$\kappa = \lambda_1/\lambda_3 \tag{27}$$

The projection of the three-dimensional eigenvalue ellipsoid into the x-y plane reveals the least observable Cartesian direction. As expected, Fig. 8 shows, in both cases, this worst direction to be perpendicular to the recent direction of travel up to that point. The slight difference in direction shown by the two bold arrows in the figure is reflected in the distribution of each of the two groups of five estimates based on five varying batches of observations. The magnitudes of the two arrows are inversely proportional to λ_3, whose magnitude is also reflected in the extent of distribution of the two sets of estimates. Because of the choice of units—radians for orientation and inches for position—the eigenvector associated with λ_1 is almost completely perpendicular to the x-y plane, that is, it is associated with estimates of vehicle orientation, ϕ. Thus λ_2 and λ_3, together with their respective eigenvectors, are indicators of the direction and magnitude of variance of the best in-plane direction (for estimates) and the worst in-plane direction.

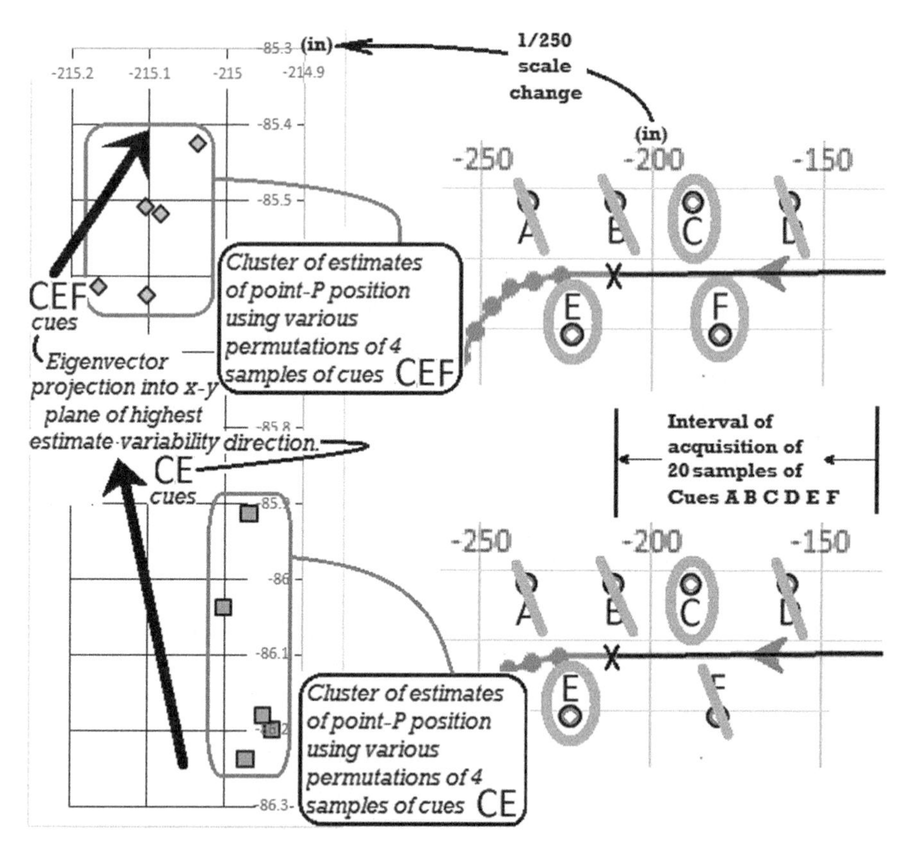

Fig. 8 Estimates of the state at the same point, applying different groups of samples of the same wall-cue combinations all acquired during the same single run—(top) involving Cues C, E, and F, and (bottom) involving only Cues C and E

5 Fixing a Through-the-Door Problem

The usefulness of the above analysis for batch selection during subsequent tracking events can perhaps best be appreciated by relating anecdotally the experience with the above trajectory: With a more or less ad-hoc selection and weighting of the samples using the most-recent 20 observations—with the caveat that no more than the most-recent 4 of these entailing any one particular cue would be applied—there was one problematic portion of path: The final doorway re-entry into the bathroom, shortly before "finish" in Fig. 6. Seemingly randomly, the vehicle would deviate laterally and sometimes hit the side of the entry rather than pass through.

Figure 9 shows the distribution of cues in the vicinity of this portion of the maneuver. As a test, a new, additional cue was added, "Cue 0," to the local group of prior cues, as shown in Fig. 9. The same index, κ, of matrix condition was calculated, using data taken in a new run, across the interval shown in Fig. 9.

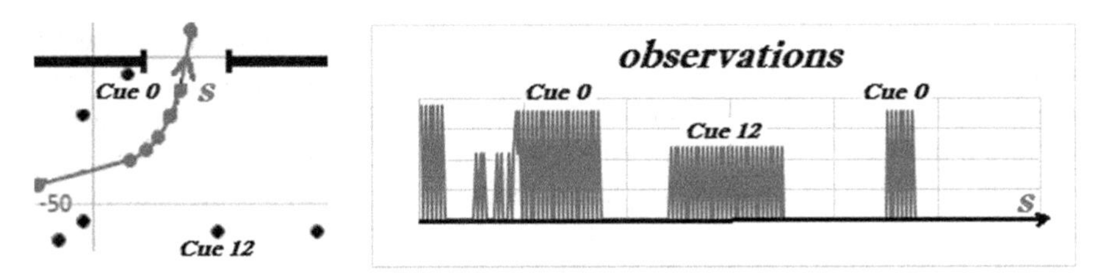

Fig. 9 Introduction of Cue 0 into the problematic region of door entry; each spike to the right is a cue observation. Note first cluster and second cluster of Cue-0 observations

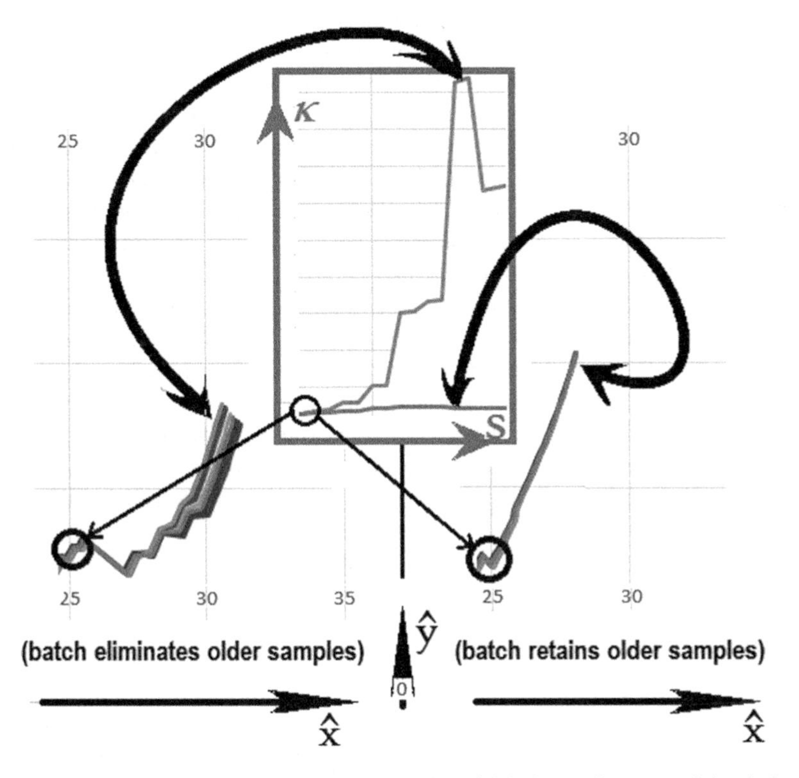

Fig. 10 Effect on 10 x-y position-estimate trajectories of deleting earlier, vs retaining in batch, "first-cluster" Cue-0 samples

Figure 9 also shows the series of observations of Cue 0 across the interval. Each peak represents a new sample. The corresponding sequence of κ, based on the aforementioned ad-hoc method for batch selection at any point along the path segment, s, and equal weighting of 20 samples, is shown in Fig. 10.

The condition number abruptly rises after four samples of Cue 0 have been acquired in the second grouping of Fig. 9; this is just prior to attempted entering through the door.

In order to interpret this peak, consider the large number of samples of Cue 12 that occurs between the two clusters of Cue-0 samples, as shown in Fig. 9. Off-line, using various groups of two of these many Cue-12 samples, the estimates leading up to the door were recalculated. Although other details of the batch were held constant, as indicated in the lower left of Fig. 10, the spike in κ coincided with an abrupt separation of the various estimates. The range of estimates at the same juncture in the trajectory varies by almost 2 inches. This is consistent with the poor repeatability that had been observed physically passing through the door.

For contrast, the identical 10 groups of Cue-12 (nonintersecting sets of 2) samples were applied to 10 new estimates of the state across the same portion. This time, the "old," group of four samples of Cue 0, from the first cluster of samples shown in Fig. 9, was retained. Earlier, older samples of other cues were dropped. The κ series changed dramatically, as indicated in the center of Fig. 10; it did not spike at all. In keeping with this result, the ten different estimates actually converged further toward each other, rather than diverging. Whereas the variability in Cue-12 membership of the batch accounted for nearly two inches of variation among the estimates absent the early Cue-0 participation, this same variability resulted in a mere quarter of an inch range with the early cues included.

It is interesting that mean estimates of the two groups are significantly different—by about 4 inches. Provided, with the inclusion and similar weighting of first-cluster samples of Cue 0, the repeatability holds up when comparing the teaching-phase estimates against repeat-phase estimates, this is not a deterrent. It is similar to the distinction among industrial, holonomic robots between repeatability and accuracy—where an order-of-magnitude difference, in favor of the former, is the norm, but where success of the repeat operation depends upon realizing the tighter repeat-phase precision.

6 Alternative Formulation

In addition to calculating, according to the equal-weighting case of Eqs. (11)–(22), $\boldsymbol{A}^T\boldsymbol{A}$, the least-squares code was used off-line to approximate, at any given juncture s along the trajectory, via finite difference, the matrix B given by

$$B_{1i} = \frac{\partial \boldsymbol{x}_A}{\partial zi} \,|\, \mathbf{x}_A \tag{28}$$

$$B_{2i} = \frac{\partial \boldsymbol{y}_A}{\partial zi} \,|\, \mathbf{x}_A \tag{29}$$

$$B_{3i} = \frac{\partial \boldsymbol{\phi}_A}{\partial zi} \,|\, \mathbf{x}_A \tag{30}$$

From Eq. (11), we see

$$\boldsymbol{B} = \left[\boldsymbol{A}^T\boldsymbol{A}\right]^{-1}\boldsymbol{A}^T \tag{31}$$

from which

$$\boldsymbol{B}^T = \boldsymbol{A}\left[\boldsymbol{A}^T\boldsymbol{A}\right]^{-T} \tag{32}$$

and, since $\boldsymbol{A}^T\boldsymbol{A}$ is symmetric,

$$\left[\boldsymbol{B}\boldsymbol{B}^T\right]^{-1} = \boldsymbol{A}^T\boldsymbol{A} \tag{33}$$

Values of κ calculated using $[\boldsymbol{B}\boldsymbol{B}^T]^{-1}$ were in practice generally very close (within 1%) of those calculated using $\boldsymbol{A}^T\boldsymbol{A}$. The exception occurred when the vehicle was making a relatively sharp turn, such as that shown in Fig. 9. It is in similar cases where it is believed the EKF differs (unfavorably) from the MHE; and in these cases the two sets of κ values can differ, as indicated in Fig. 11, by 50%. We use the (generally more conservative, higher valued) case where κ is determined from finite differencing.

The overall lesson is that a necessary condition for repeatability, for any given formulation of wall-cue location, batch collection, and trajectory plan, is a low condition number calculated off-line, based upon teaching-phase data, using the finite-difference-based $[\boldsymbol{B}\boldsymbol{B}^T]^{-1}$.

Sufficient conditions for repeatability extend to the matters of control/tracking accuracy, assurance of acquisition of requisite cue indications, and reasonably confined wheel slip. One advantage of the wheeled device, as distinct from a flying device, is that all three of these things can normally be ensured by way of slowing the speed of the vehicle, which does not otherwise diminish the above algorithms. Regarding the need for good tracking: dynamics-induced deviation from the reference path, provided the estimates are accurate, is always "known" within the program, and will almost always tend toward zero as speed is lowered where this is detected. Likewise, wheel slip, unless external forces are present, is controlled by slower speeds on curves and reduced acceleration, both of which can be controlled automatically. Rolling slip is now well-detected during aircraft landing and with autos as with anti-lock brakes.

If, during tracking, early efforts to detect a key cue in a key region of the repeated path fail, then there is also the prospect of reducing speed in order to "try harder" (e.g., apply transported lighting or switch to more sensitive cue-detection software.) There is even the prospect of reversing direction along the path, briefly, and "trying again."

A remaining condition for sufficiency is a "reasonably" accurate model. This would include geometric aspects of the chair and cameras, locally (regionally) con-

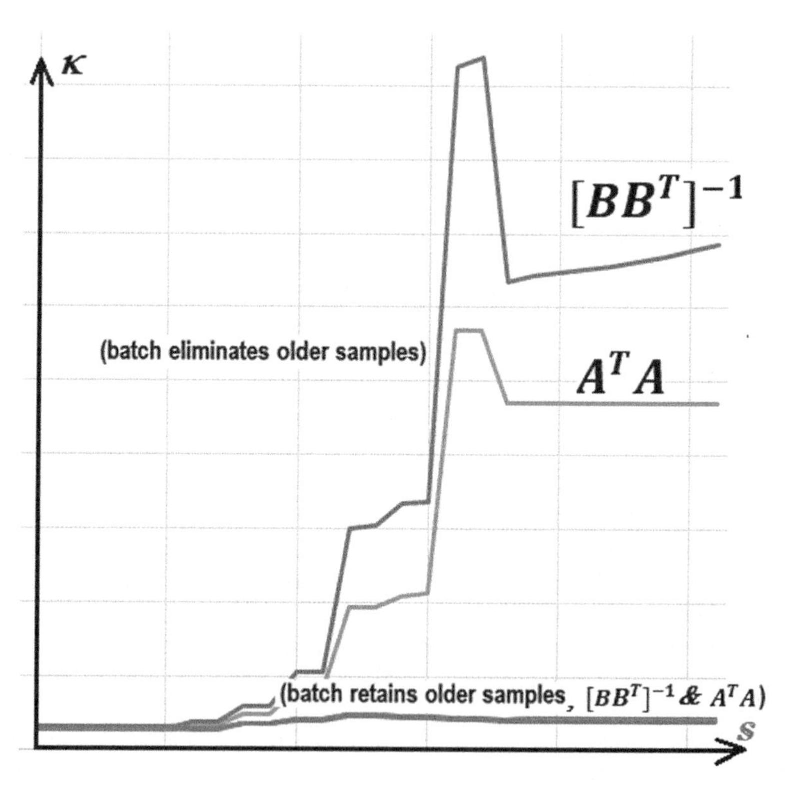

Fig. 11 Alternate formulations of condition number for the interval shown in Fig. 9. Shown for each of 2 batch formulations

sistent measurements/entries of cue locations, and well-calibrated intrinsic camera parameters. Importantly, *global* consistency of the cue locations is unnecessary if, locally, cue locations of proximate cues are accurate, one in relation to another. In this last point, there is surprising insensitivity to a random corruption, corruption by as much as an inch, of just one of the cues even in a *local* cluster of (say) 4 cues used in the estimates within a region of a taught path. Depending upon a variety of factors, the present algorithm compensates for entry inaccuracy due to the use of the *same* corrupted location-entry for *both* teaching and tracking. The effect on tracking tends to be undetectable even for an inch of geometric incompatibility with respect to the other 3 cues.

7 Summary and Conclusions

The prospect of using teach/repeat for a nonholonomic wheelchair in a geometrically imprecise, but permanent, setting is set forth in terms of achieving a high level of repeatability. This repeatability criterion has a counterpart in industrial holonomic robots. An important necessary condition for this goal is shown herein using data from an experimental, home system with high-precision requirements and wall fiducials; it uses locally accurate (commensurate) pre-entered wall positions but makes no effort to produce global accuracy. It is shown that, during a setup phase, there is a good prospect of testing taught trajectories off-line principally by means of condition-number calculation, in order to determine whether differing batches of observations, drawn from the same run, will produce very nearly identical results of the corresponding sequence of vehicle poses. This is a first step in the direction of creating useful, real-world autonomy.

Earlier, EKF-based efforts toward this objective [2] suffered from inadequate estimation reliability and accuracy. That is one reason why the approach was not pursued beyond some trials at the Hines VA Hospital in Hines, IL. The other reason had to do with optimism for Simultaneous Localization and Mapping (SLAM) [10]. Developed vigorously by the robotics community, SLAM might have obviated the need for the preplaced, otherwise-unnecessary wall cues as well as the need for trajectory teaching episodes, during set-up, to connect between every pose-of-departure/pose-of-arrival combination as may be needed by the severely disabled veteran.

Faster computing, more precise and reliable estimation due to MHE per the present approach (https://www.youtube.com/watch?v=mZ0cadG2ahM&feature=youtu.be), and lack of emergence of SLAM, encourages further consideration of teach/repeat.

Additional, currently unmet needs for autonomous vehicles might be developed relatively easily and robustly using the same approach, particularly if/as ancillary obstacle avoidance—exploiting proximity-sensor records from previous, obstacle-free runs of the same trajectory—works well. While one urgent application may be the severely disabled veteran, other wheelchair-bound users, including the blind, could benefit. Nonredundant, comprehensive floor maintenance, to replace bounded random motion, is also a good possibility. Also possible are surveillance and delivery within a given building or vessel.

Acknowledgment This research was supported in part by the US Department of Veterans Affairs, and in part by the Naval Center for Applied Research in Artificial Intelligence, US Office of Naval Research.

References

1. L. Fehr, E. Langbein, S. Skaar, Adequacy of power wheelchair control interfaces for persons with severe disabilities: a clinical survey. J. Rehabil. Res. Dev. **37**(3), 253–260 (2000)
2. E. Baumgartner, S. Skaar, An autonomous vision-based mobile robot. IEEE Trans. Automat. Contr. **39**(3), 493–502 (1994)
3. M. Perrollaz, S. Khorbotly, A. Cool, J. Yoder, E. Baumgartner, Teachless teach-repeat: toward vision-based programming of industrial robots, in *Proceedings of IEEE International Conference on Robotics and Automation*, St Paul, MN, USA (May 2012)

4. T. Whitworth, Fixturing for automated welding in automotive—repeatability, access, and protection are key. MetalForming **48**(2) (2014)
5. H. Goldstein, *Classical Mechanics*, 3rd edn. (Addison Wesley, 1980), p. 16
6. A. Gelb (ed.), *Applied Optimal Estimation* (MIT Press), 1974
7. E. Haseltine, J. Rawlings, Critical evaluation of extended kalman filtering and moving-horizon estimation. Ind. Eng. Chem. Res. **44**(8), 2451–2460
8. J.L. Junkins, C. White, J. Turner, Star pattern recognition for real-time attitude determination. J. Astronaut. Sci. **25**, 251–270 (1977)
9. D. Belsley, E. Kuh, R. Welsch, *The Condition Number, Regression Diagnostics: Identifying Influential Data and Sources of Collinearity* (John Wiley & Sons, New York), pp. 100–104
10. C. Cadena, L. Carlone, H. Carrillo, Y. Latif, D. Scaramuzza, R. Neira, J. Leonard, Past, present, and future of simultaneous localization and mapping: toward the robust-perception age. IEEE Trans. Robot. **32**(6), 1309–1336

Exact Solutions to the Spline Equations

Anthony A. Ruffa and Bourama Toni

1 Introduction

Spline curves are mathematically described by piecewise polynomials. They provide an interface to design and control the shape of complex curves and surfaces.

In particular, *cubic splines* are a well-used tool[1] for curve fitting for a given sequence of data points called *control points* [1]. Indeed, cubic polynomials are the lowest-degree polynomial to support an inflection, commonly present in most complex curves; they are also well-behaved numerically, yielding usually smooth curves. High-degree polynomials tend to exhibit greater sensitivity to the positions of the control points. The process of cubic curve fitting consists of constructing a complex curve with a high number of inflection points and several pieces of cubic curves $f_k(x) = y_k = a_k + b_k x + c_k x^2 + d_k x^3$ on each segment $[x_k, x_{k+1}]$ such that

[1] Applications are mostly in CAD (computer-assisted design), CAM (computer-assisted manufacturing), and computer graphics systems when an operator wants to draw a smooth curve through data points not subject to error.

A. A. Ruffa (✉)
Naval Undersea Warfare Center, Newport, RI, USA
e-mail: anthony.ruffa@navy.mil

B. Toni
Department of Mathematics, Howard University, Washington, DC, USA
e-mail: bourama.toni@howard.edu

A. A. Ruffa, B. Toni (eds.), *Advanced Research in Naval Engineering*, STEAM-H: Science, Technology, Engineering, Agriculture, Mathematics & Health, https://doi.org/10.1007/978-3-319-95117-1_7

they satisfy the so-called C^r−continuity, $r = 0, 1, 2$.[2] That is, each curve passes through its control points, the slopes and the curvatures match where the curves join. Or, equivalently

1. $f_k(x_k) = y_k$. $f_k(x_{k+1}) = y_{k+1}$.
2. $f_k'(x_{k+1}) = f_{k+1}'(x_{k+1})$
3. $f_k''(x_{k+1}) = f_{k+1}''(x_{k+1})$.

Note here that the second derivatives $f_k''(x) = 2c_k + 6d_k x$ are linear functions. Therefore, inflection points will appear if and only if the second derivative y_k'' and y_{k+1}'' have opposite sign leading to additional curvature conditions to avoid *unwanted* inflection points, the so-called *extraneous inflection points,* indicating a change of concavity where no such reversal exists in the actual physical problem being modeled. As for the physical problem for design ship's hull and automobile body, the objective is to produce a smooth geometric outline connecting the knots under specific curvature conditions. In other words, sometimes certain regions require a positive second derivative when the cubic spline is generating a negative one, creating an unwanted inflection point; and in such a case, the usual cubic spline does not allow any correction, thus the need of a cubic spline endowed with a parameter which can be varied to generate the curvature conditions leading to the correct smooth geometric outline; this is provided by the so-called *hyperbolic cubic spline* or cubic spline under tension described in Sect. 4; the tension factor acts as a correcting parameter that can be changed appropriately to remove all unwanted *extraneous* inflection points, source of undesirable oscillations in certain regions. As such, the *spline under tension* represents better the early draftsman's techniques.

It is common to give the slope at both ends of the curve leading to zero second derivative, for the so-called *natural cubic splines.* However, cubic splines can oscillate within segments in the region of a discontinuity, including discontinuity in slopes. For the case of equal knot spacing, the cubic spline equations form a symmetric tridiagonal Toeplitz system. Such systems have an exact solution [3], which can support a more systematic investigation into the mechanisms governing the oscillatory response.

Fitting a spline to a data set approximating a step function illustrates the oscillatory response that can occur in the region of a discontinuity, and will be the primary focus here. Splines under tension [2] attempt to simulate a spline as a cable under tension, which also leads to a symmetric tridiagonal Toeplitz system for the case of equal knot spacing. Here, the tension can be increased to attenuate the oscillations over a shorter length scale.

This chapter is organized as follows. Section 2 develops the exact solution for a cubic spline with equal knot spacing and computes a curve fit for a single nonzero data point. The superposition of such solutions yields a curve fit for an arbitrary data set (with an emphasis on data sets approximating a step function).

[2]Draftmen realize a smooth interpolation curve with a long flexible beam, a spline, constrained to pass by all given points, outlining the deflection curve using heavy objects (*the drawing dogs*).

Section 3 focuses on the effect of end conditions for the cubic spline. Section 4 focuses on splines under tension and their potential for mitigating oscillations. The use of imaginary tension is also explored as an approach to remove the high-amplitude oscillations in the region of a discontinuity (at the expense of introducing low-amplitude oscillations throughout the entire curve fit). A "composite spline" provides a means to confine the remaining oscillations to arbitrarily defined regions. Finally, the conclusions are presented in Sect. 5.

2 Cubic Splines

We briefly recall the mathematical spline[3] as follows:

Definition 1 Given $a = x_0 < x_1 < \ldots < t_N = b$, a function S is called a *spline of degree* $n \geq 1$ or *n-spline* with the respect the so-called *knots* x_k if it satisfies the following:

1. S is a polynomial of degree at most n over the interval $[x_k, x_{k+1}]$,
2. $S \in C^{n-1}[a, b]$ (*smoothness condition*)

Note that possessing continuous first- and higher-order derivative makes a piecewise polynomial even smoother.

A *natural cubic spline*[4] for n equally spaced knots having data values y_k (for $1 \leq k \leq n$) involves a curve fit Y_k to each segment (which is assumed to have a unit length) as follows [1]:

$$Y_k(t) = a_k + b_k t + c_k t^2 + d_k t^3; \tag{2.1}$$

where

$$0 \leq t \leq 1;$$

$$Y_k(0) = a_k = y_k;\ Y_k(1) = y_{k+1} = a_k + b_k + c_k + d_k$$

$$Y_k'(0) = b_k = D_k;\ Y_k'(1) = D_{k+1} = b_k + 2c_k + 3d_k$$

[3]Most of the early studies on spline were done by I.J. Schoenberg (1903–1990), often referred to as the *father of splines*.

[4]This is the curve generated by forcing a flexible elastic rod into the data points but letting the slope at the ends be free to adjust to positions that minimizes the oscillatory behavior of the curve.

Solving these four equations yields

$$c_k = 3(y_{k+1} - y_k) - 2D_k - D_{k+1};$$

$$d_k = 2(y_k - y_{k+1}) + D_k + D_{k+1}. \tag{2.2}$$

Here, D_k is the first spatial derivative at knot k. The continuity of the first derivative means that the resulting piecewise curve has no sharp corners, while the continuity of the second derivative defines the radius of curvature at each point. The process is called *Hermite Interpolation.* Imposing the requirement of continuous first and second spatial derivatives at each knot, as well as a zero second derivative at both endpoints leads first to

$$D_{k-1} + 4D_k + D_{k+1} = 3(y_{k+1} - y_{k-1}), \tag{2.3}$$

and ultimately to the following system of equations [1]:

$$\begin{bmatrix} 2 & 1 & & & & & \\ 1 & 4 & 1 & & & & \\ & 1 & 4 & 1 & & & \\ \ddots & \ddots & \ddots & \ddots & \ddots & \ddots & \ddots \\ & & & 1 & 4 & 1 & \\ & & & & 1 & 4 & 1 \\ & & & & & 1 & 2 \end{bmatrix} \begin{bmatrix} D_1 \\ D_2 \\ D_3 \\ \vdots \\ D_{n-2} \\ D_{n-1} \\ D_n \end{bmatrix} = \begin{bmatrix} 3(y_2 - y_1) \\ 3(y_3 - y_1) \\ 3(y_4 - y_2) \\ \vdots \\ 3(y_{n-1} - y_{n-3}) \\ 3(y_n - y_{n-2}) \\ 3(y_n - y_{n-1}) \end{bmatrix}. \tag{2.4}$$

Equation (2.4) represents a symmetric tridiagonal Toeplitz system and thus has an exact solution [3]. The row structure is [1, 4, 1], except for the first and last rows. The exact solution leads to the solution to Eq. (2.1).

As a shorthand, we denote (2.4) as $A_{jk}D_k = B_j$ (omitting the summation symbol). The solution is derived from the following lemma:

Lemma 1 *The system of equations*

$$A_{jk}X_k^{(m)} = \delta_{jm} \tag{2.5}$$

where δ_{jm} is the Kronecker delta, admits a unique exact solution for each value of $m \in [1, n]$.

Proof First seeking solutions of the form $e^{\gamma k}$ [3] the Toeplitz row structure [1, 4, 1] then leads to $e^{\gamma k} + 4e^{\gamma(k+1)} + e^{\gamma(k+2)} = 0$, so that $e^{\gamma} = -2 \pm \sqrt{3}$, or $\gamma_{1,2} \cong \pm 1.3170 + \pi i$.

Denoting $\alpha = \mathrm{Re}(\gamma_1)$, the following solutions can be developed:

$$E_k^{(m)} = e^{\alpha(k-m)} \cos \pi(k-m); \tag{2.6}$$

$$F_k^{(m)} = e^{-\alpha(k-m)} \cos \pi(k-m) = \frac{1}{E_k^{(m)}}; \tag{2.7}$$

$$G_k = e^{-\alpha k} \cos \pi k; \tag{2.8}$$

$$H_k = e^{\alpha(k-n)} \cos \pi k. \tag{2.9}$$

For $2 \leq m \leq n-1$, the solutions (2.6)–(2.9) can be superimposed as follows:

$$\hat{X}_k^{(m)} = E_k^{(m)} + C_1^{(m)} G_k + C_2^{(m)} H_k;\ k < m; \tag{2.10}$$

$$\hat{X}_k^{(m)} = F_k^{(m)} + C_1^{(m)} G_k + C_2^{(m)} H_k;\ k \geq m. \tag{2.11}$$

The constants $C_1^{(m)}$ and $C_2^{(m)}$ result from the solution of the following equations:

$$C_1^{(m)}(2G_1 + G_2) + C_2^{(m)}(2H_1 + H_2) + (2E_1^{(m)} + E_2^{(m)}) = 0; \tag{2.12}$$

$$C_1^{(m)}(G_{n-1} + 2G_n) + C_2^{(m)}(H_{n-1} + 2H_n) + (F_{n-1}^{(m)} + 2F_n^{(m)}) = 0. \tag{2.13}$$

The solution to $A_{jk} X_k^{(m)} = \delta_{jm}$ for $2 \leq m \leq n-1$ is then as follows:

$$X_k^{(m)} = \frac{\hat{X}_k^{(m)}}{C_3^{(m)}}; \tag{2.14}$$

where

$$C_3^{(m)} = \hat{X}_{m-1}^{(m)} + 4\hat{X}_m^{(m)} + \hat{X}_{m+1}^{(m)};\ 2 \leq m \leq n-1. \tag{2.15}$$

The endpoints correspond, respectively, to $m = 1$ and $m = n$ as follows:

For $m = 1$, the solution is

$$X_k^{(1)} = \frac{F_k^{(1)}}{C_3^{(1)}}, \tag{2.16}$$

where

$$C_3^{(1)} = 2F_1^{(1)} + F_2^{(1)}. \tag{2.17}$$

Finally, for $m = n$, the solution is

$$X_k^{(n)} = \frac{E_k^{(n)}}{C_3^{(n)}}, \tag{2.18}$$

where

$$C_3^{(n)} = E_{n-1}^{(n)} + 2E_n^{(n)}. \tag{2.19}$$

Hence, the claim.

We then prove

Theorem 1 *The exact solution to Eq. (2.3), and hence the solution to Eq. (2.1),is obtained as*

$$D_k = 3\sum_{j=2}^{n-1} X_k^{(j)} \cdot (y_{j+1} - y_{j-1}) + 3X_k^{(1)} \cdot (y_2 - y_1) + 3X_k^{(n)} \cdot (y_n - y_{n-1}). \tag{2.20}$$

where the term $X_k^{(m)}$, $1 \leq m \leq n$, *is given in the above lemma.*

Proof The exact solution is derived from the superposition of the solutions to the n systems $A_{jk}X_k^{(m)} = \delta_{jm}$, each weighted by the appropriate right-hand side (RHS) vector term in (2.1), leading to the expression for D_k as claimed in the theorem.

Remark 1 From the expression of the solution

$$D_k = 3\sum_{j=2}^{n-1} X_k^{(j)} \cdot (y_{j+1} - y_{j-1}) + 3X_k^{(1)} \cdot (y_2 - y_1) + 3X_k^{(n)} \cdot (y_n - y_{n-1}), \tag{2.21}$$

we note that for interior knots, $X_k^{(m)}$ grows exponentially for $k < m$ and decays exponentially for $k > m$. The $\cos \pi k$ term in (2.6), (2.7), (2.8), and (2.9) results from the $[1, 4, 1]$ row structure of A_{jk} in (2.4).[5] The solution $X_k^{(m)}$ (Fig. 1) has compact support, and every other term changes sign.

To illustrate, we take $m = 100$ and $n = 200$.

[5]The eigenvalues for $[1, 4, 1]$ tridiagonal symmetric Toeplitz matrix are given by $\lambda_k = 4 + 2\cos\frac{k\pi}{n+1}$.

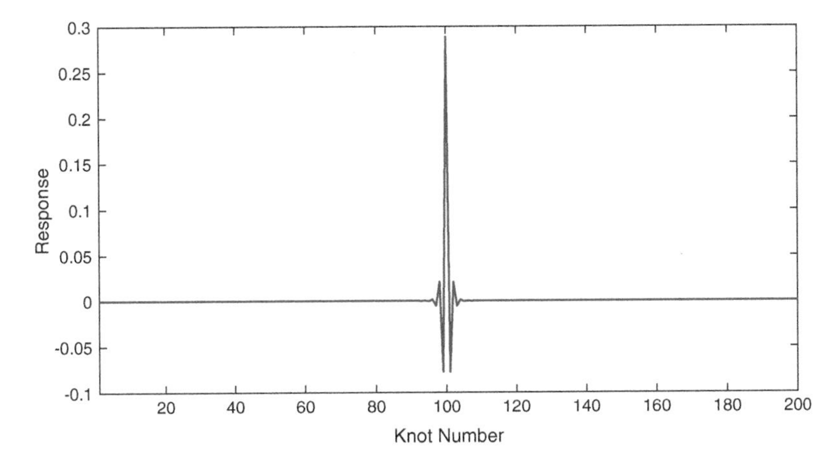

Fig. 1 The solution to $A_{jk}X_k^{(m)} = \delta_{jm}$ for $m = 100$ and $n = 200$

We consider next the cases of a data set that approximates, respectively, a step function and a sine wave; however, most of the analysis henceforth is done graphically, and illustrates how oscillations cancel out by superposition of solutions.

2.1 Approximating a Step Function

The RHS vector B_j for a data set that approximates a step function (Fig. 2) has nonzero data points at two adjacent knots. The derivative vector D_k (Fig. 3) is generated by superimposing the solutions for two systems, each with an RHS vector having a single nonzero data point.

A curve fit of an n-knot data set without discontinuities can be broken down into n systems, each representing a curve fit to a single nonzero data point. Figure 4 shows one such curve fit. When the n single-point curve fits are superimposed, the oscillations cancel.

The cancellation effect also occurs in curve fits of data sets approximating step functions. For example, the curve fits in Figs. 5 and 6 oscillate in the region of the discontinuities; however, superimposing them (Fig. 7) yields a curve fit of a straight line, and the oscillations cancel.

2.2 Approximating a Sine Wave

The curve fit to an n-knot data set that approximates a sine wave (Fig. 8) can also be solved as n separate cubic splines, each with only one nonzero data point. Figure 9 shows the curve fit to a single nonzero data point (at $k = 50$), which is identical

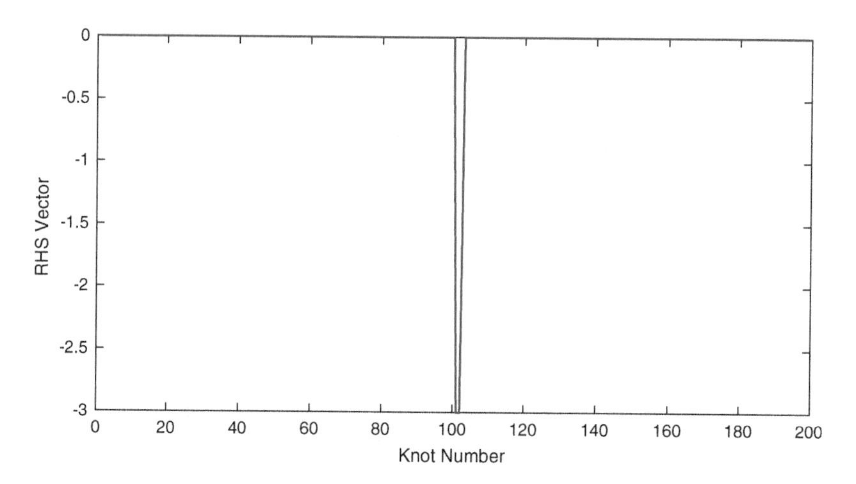

Fig. 2 The RHS vector B_j for a data set approximating a step function, i.e., $y_k = 1$ for $1 \le k \le 100$, and $y_k = 0$ otherwise

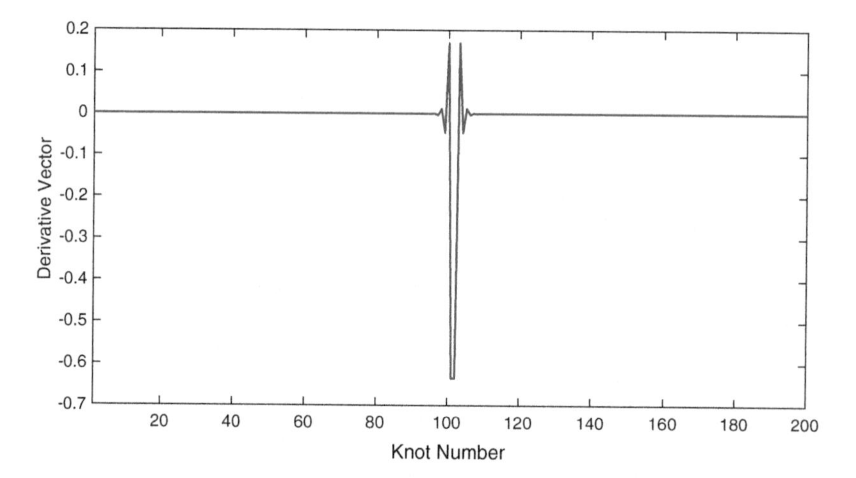

Fig. 3 The derivative vector D_k corresponding to a data set approximating a step function

to the curve fit in Fig. 4, except for differences in the magnitude and knot number. When n such curve fits are superimposed, the oscillations cancel (Fig. 8). Figure 10 shows the superposition of $n/2$ such curve fits taken from the central region of the data set. The oscillations cancel except at the edges. Superimposing the curve fit to the remaining nonzero data points (Fig. 11) with that of Fig. 10 recovers the curve fit of Fig. 8.

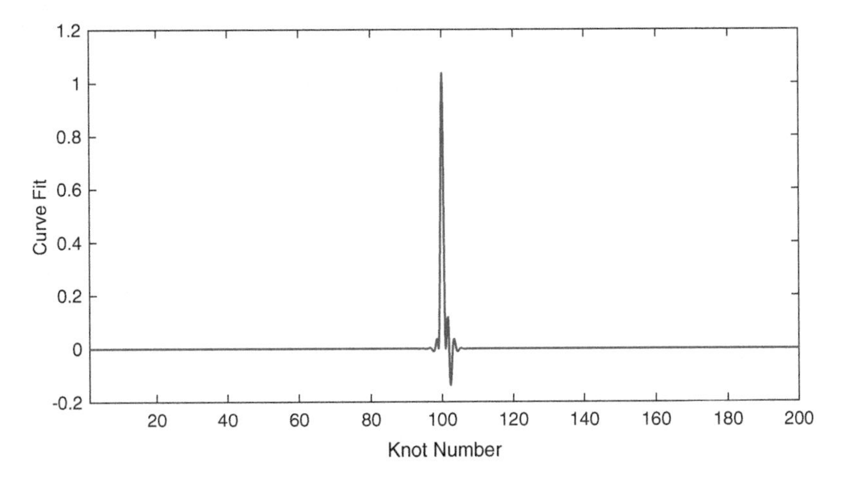

Fig. 4 The curve fit for a single nonzero data point at $k = 100$

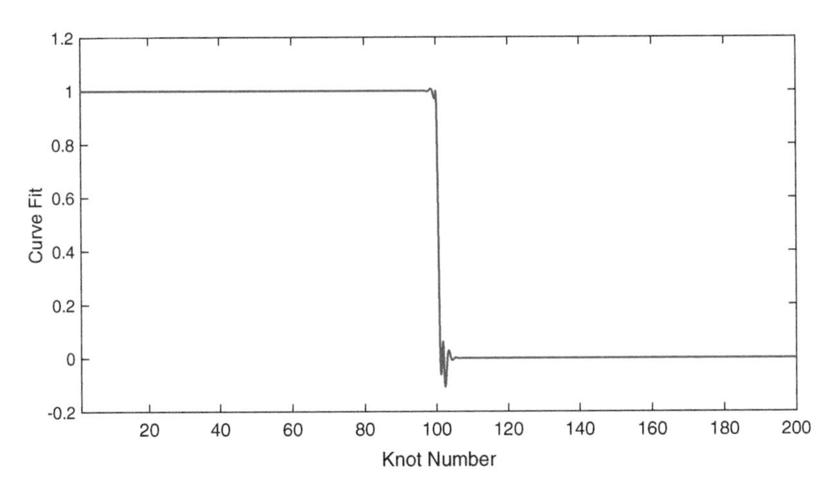

Fig. 5 The curve fit corresponding to a data set approximating the step function defined by $y_k = 1$ for $1 \leq k \leq 100$ and $y_k = 0$ otherwise

3 End Conditions

Endpoint constraints lead to different types of cubic splines, from the *clamped spline* with specified first derivative boundary conditions, to *parabolically terminated spline* with zero third derivative on the first and last segment, to *curvature-adjusted spline* with the second derivative boundary conditions specified, to the *natural cubic spline* with zero end second derivatives, which is our focus here. The relevant feature of the resulting spline is the minimum of the oscillatory behavior they possess.

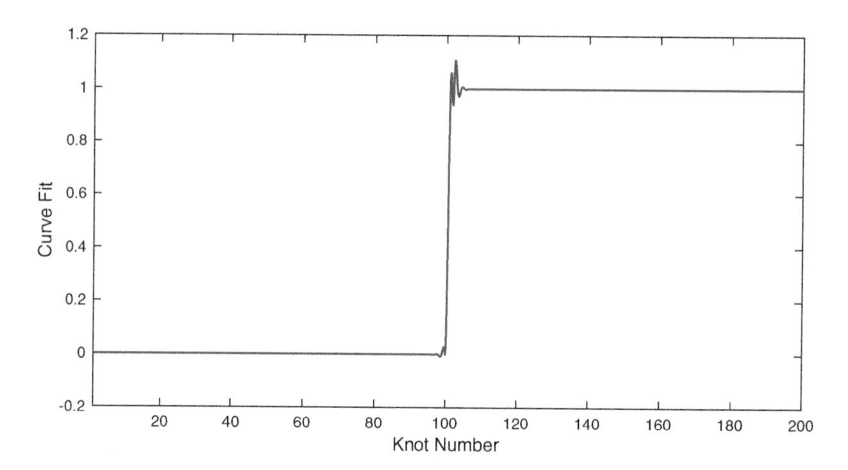

Fig. 6 The curve fit corresponding to a data set approximating the step function defined by $y_k = 1$ for $100 < k \leq n$ and $y_k = 0$ otherwise

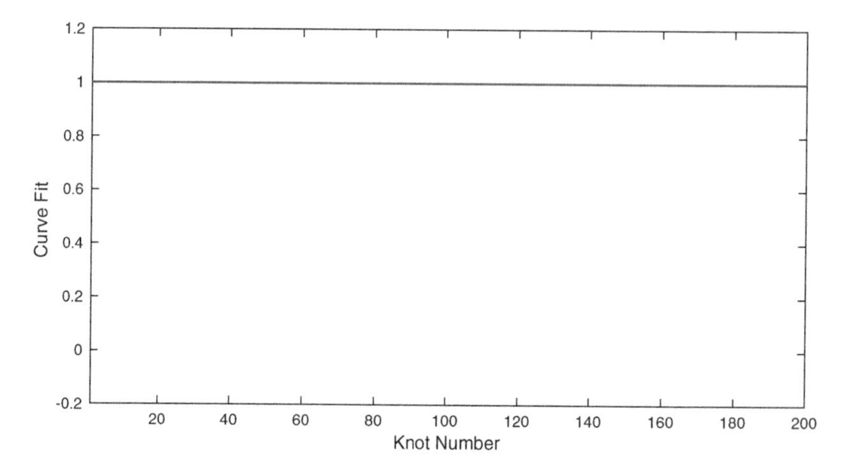

Fig. 7 The superposition of the curve fits from Figs. 5 and 6. Note that the oscillations cancel

For small values of n, the end conditions can affect the entire system [1]. However, for large values of n, the end conditions have little or no influence outside of a small region. This can be inferred from the curve fit to a single nonzero data point at the left boundary (Fig. 12), which has compact support.

A natural cubic spline (in which the second spatial derivative is set to zero at the ends) is a common end condition. However, other end conditions are possible, e.g., the first spatial derivative at each end can be set to zero. The difference between curve fits based on these two end conditions (Fig. 13) also has compact support.

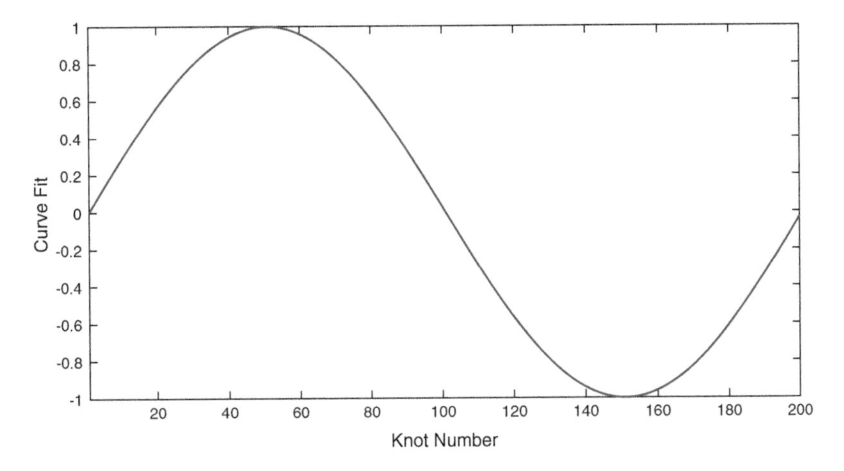

Fig. 8 The curve fit for a data set approximating a sine wave ($n = 200$)

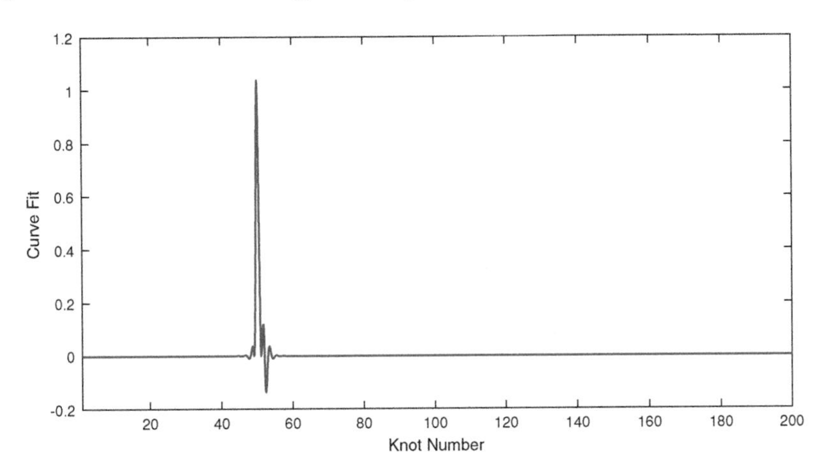

Fig. 9 The curve fit for one nonzero data point taken from the sine wave data set

4 Splines Under Tension

The standard cubic splines oftentimes exhibit unwanted (*extraneous*) inflection points, with no correcting parameter to realize a smooth geometric outline. The interpolation cubic curve has generally at most one inflection point on each interval $[x_k, x_{k+1}]$. Schweikert unwanted inflection points are defined as follows:

Definition 2 Let the second difference at x_k be given by

$$d_k = (y_{k+1} - y_k) - (y_k - y_{k-1}) = y_{k+1} - 2y_k + y_{k-1}. \tag{4.1}$$

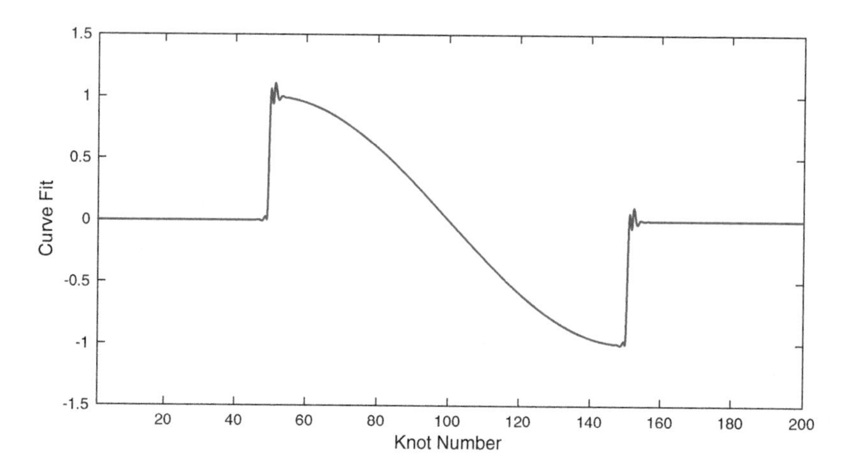

Fig. 10 The curve fit for 100 nonzero data points taken from the central region of the sine wave data set

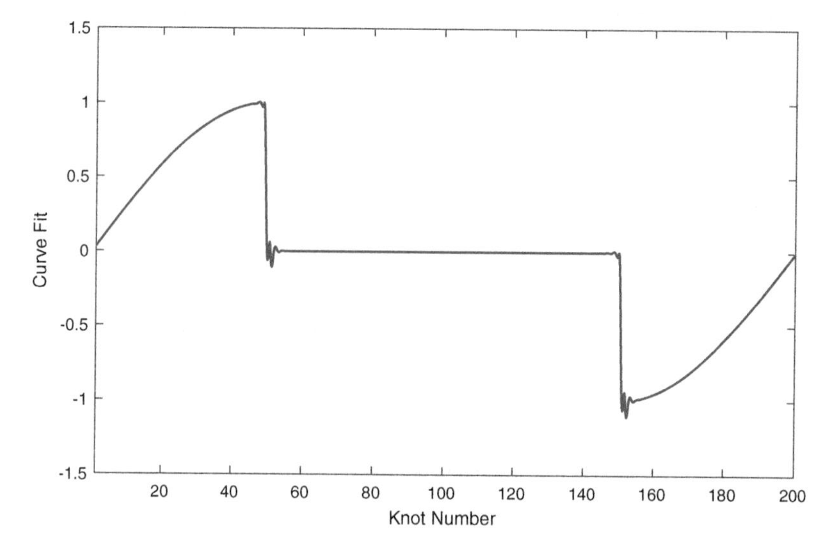

Fig. 11 The curve fit for the remaining nonzero data points taken from the sine wave data set. Superimposing this curve fit with that in Fig. 10 recovers the curve fit in Fig. 8

An inflection point, if any, is *extraneous* if d_k and d_{k+1} have the same sign.

Schweikert, who first introduced the splines under tension to eliminate these extraneous inflection points, proved in [4]:

Theorem 2 *The interpolation cubic spline has no extraneous inflection points if and only* y''_k *and* d_i *have the same sign.*

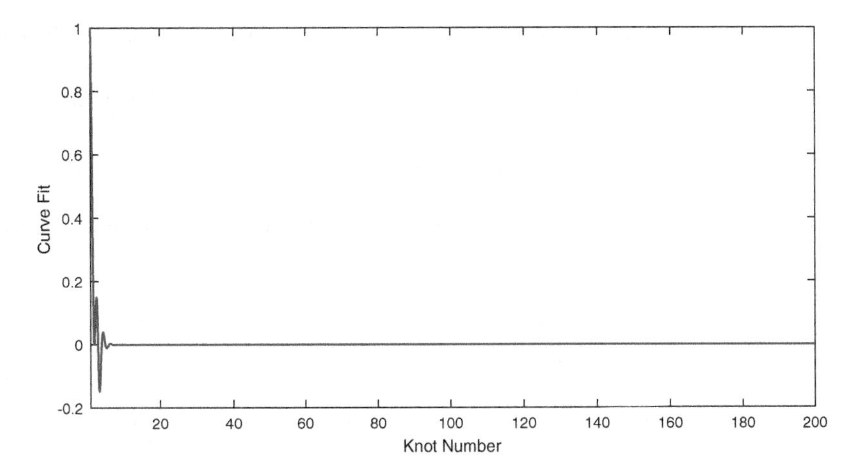

Fig. 12 The curve fit to a single nonzero data point at the first knot

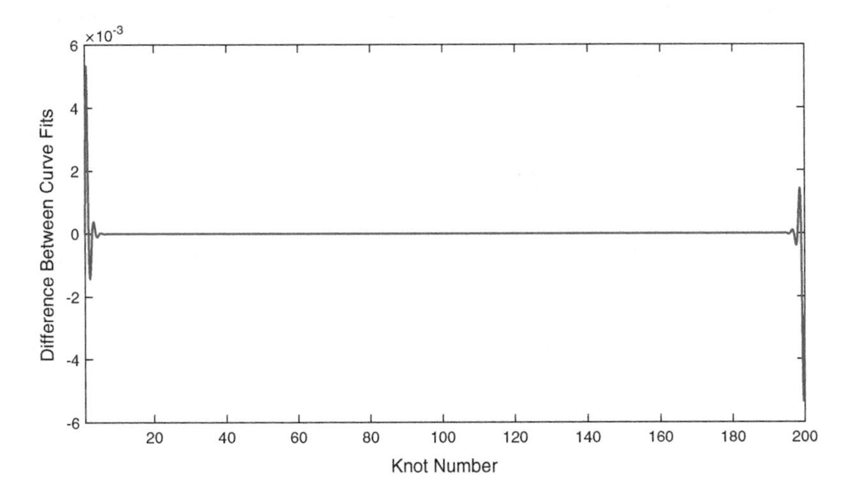

Fig. 13 The difference in curve fits for a data set approximating a sine wave, in the first case approximated by a natural cubic spline, and in the second case approximated by a cubic spline in which the first spatial derivative at each endpoint is set to zero

These additional curvature conditions are not always met by the standard cubic splines, prompting the need of a different type of interpolation curve, the *spline under tension,* also called *hyperbolic cubic spline*, in which a tension factor is added as a mathematically expediency (parameter) to be varied appropriately to remove all oscillations usually linked to *extraneous* inflection points.

Splines under tension [2] attempt to minimize the oscillations in the region of a discontinuity. For the case of equal knot spacing, splines under tension form a symmetric tridiagonal Toeplitz system. An exact solution can be developed following a similar approach to that for cubic splines. Over the interval $[x_k, x_{k+1}]$, the spline-under-tension curve fit $f(x)$ is expressed as the linear combination

$$f(x) = c_1 + c_2 + c_3 \sinh \sigma x + c_4 \cosh \sigma x$$

of the functions $1, x, \sinh \sigma x, \cosh \sigma x$ and, under the smoothness conditions, as follows [2]:

$$\begin{aligned} f(x) = f''(x_k)\frac{\sinh \sigma(x_{k+1} - x)}{\sigma^2 \sinh(\sigma h)} + \frac{(y_k\sigma^2 - f''(x_k))(x_{k+1} - x)}{\sigma^2 h} \\ + f''(x_{k+1})\frac{\sinh \sigma(x - x_k)}{\sigma^2 \sinh(\sigma h)} + \frac{(y_{k+1}\sigma^2 - f''(x_{k+1}))(x - x_k)}{\sigma^2 h}. \end{aligned} \tag{4.2}$$

Here, σ is the tension and h is the distance between knots ($h = 1$ for the examples considered here). The system of equations for $2 \le k \le n - 1$ is as follows [2]:

$$\begin{aligned} \left(\frac{1}{\sigma^2 h} - \frac{1}{\sigma \sinh \sigma h}\right) f''(x_{k-1}) + 2\left(\frac{\cosh \sigma h}{\sigma \sinh \sigma h} - \frac{1}{\sigma^2 h}\right) f''(x_k) \\ + \left(\frac{1}{\sigma^2 h} - \frac{1}{\sigma \sinh \sigma h}\right) f''(x_{k+1}) = \frac{y_{k+1} - 2y_i + y_{k-1}}{h}. \end{aligned} \tag{4.3}$$

Remark 2 Let $\eta = \frac{\sigma \cosh \sigma - \sinh \sigma}{\sinh \sigma - \sigma}$, the usual requirements for slopes and curvatures yield

$$D_{k-1} + 2\eta D_k + D_{k+1} = (\eta + 1)(y_{k+1} - y_{k-1}) \tag{4.4}$$

We also have $\eta \to 2$ for $\sigma \to 0$. Moreover, on each interval $[x_i, x_{i+1}]$ f_i'' is not linear anymore and given by

$$f_i''(x) = y_i'' \frac{\sinh \sigma(x_{k+1} - x)}{\sinh \sigma} + y_{i+1}'' \frac{\sinh \sigma(x - x_i)}{\sinh \sigma} \tag{4.5}$$

The corresponding row structure is obtained as $[1, 2\eta, 1]$ approaching $[1, 4, 1]$ as the tension approaches zero.

For $k = 1$ and $k = n$, the equations are $f''(x_1) = 0$ and $f''(x_n) = 0$, respectively. Equations (4.2) and (4.3), along with the equations at the endpoints, represent a symmetric tridiagonal Toeplitz system with $f''(x_k)$ as the unknown.

The ratio r of the diagonal term and the off-diagonal term (Fig. 14) approaches 4 as $\sigma \to 0$, so that the row structure $[1, r, 1]$ approaches that of the cubic spline system, i.e., $[1, 4, 1]$. The solutions take the form $e^{\pm\alpha(k-m)} \cos \pi(k-m)$. Increasing σ increases r and by extension, α, so that the oscillations decay over a shorter length

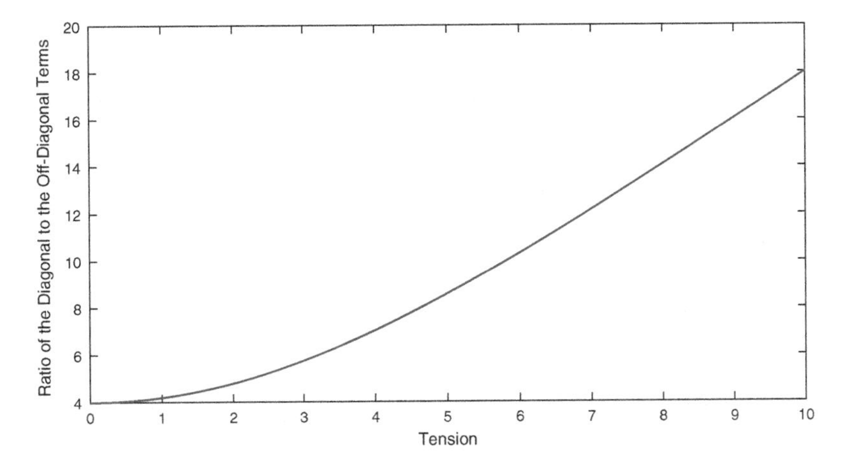

Fig. 14 The ratio r of the diagonal term and the off-diagonal term as a function of tension

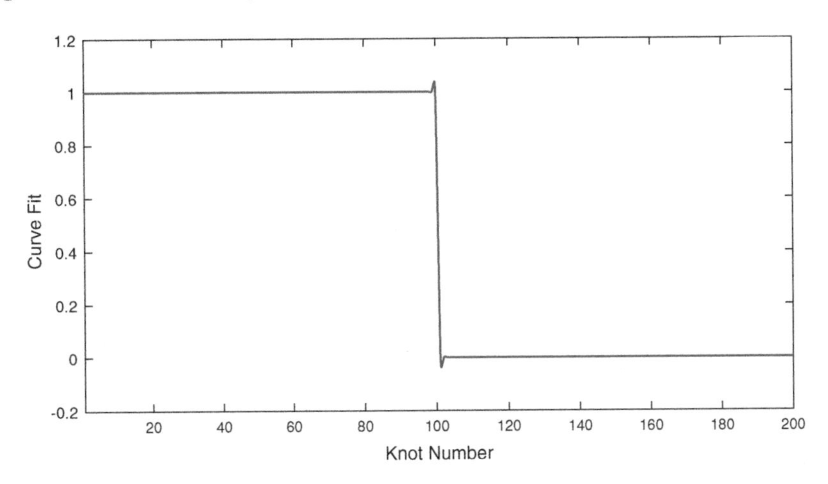

Fig. 15 A curve fit to the step function with a spline under tension for $\sigma = 10$. Note that the oscillations decay over a shorter length scale in the region of the step discontinuity relative to Fig. 5

scale (e.g., see Fig. 15). For example, when $\sigma = 1$, $\alpha \cong 1.3727$, but when $\sigma = 10$, $\alpha \cong 2.8882$. Figure 16 shows the curve fit to a single nonzero data point for $\sigma = 10$.

4.1 Splines Under Imaginary Tension

Equations (4.2) and (4.3) are unchanged when σ is replaced with $-\sigma$, indicating that σ acts differently on a spline than tension does on a cable. Another difference becomes evident when the tension becomes imaginary (since imaginary cable

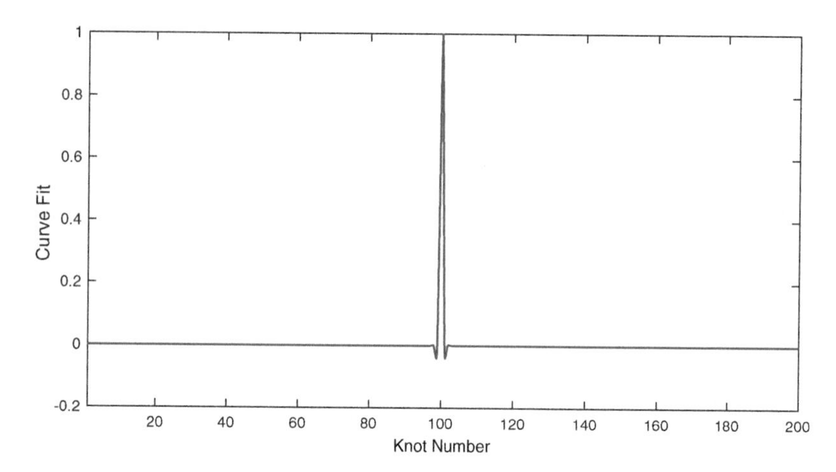

Fig. 16 A curve fit to a single data nonzero data point with a spline under tension for $\sigma = 10$. Note that the oscillations decay over a shorter length scale relative to Fig. 4

tension is impossible). Imaginary values for σ do not lead to imaginary terms in (4.2) or (4.3), and they can generate a non-oscillatory solution for values of σ resulting in a row structure $[1, r, 1]$, in which $r < 0$.

Figure 17 shows the ratio of the diagonal to the off-diagonal terms as a function of the scaled imaginary tension, i.e., $\hat{\sigma} = \mathrm{Im}(\sigma)/2\pi$. Although $r < 0$ for some values of σ, the solutions are not damped, since $r \geq -2$. Furthermore, a limiting process must be performed to obtain a system having a row structure of $[1, -2, 1]$ for $\sigma \to 2\pi i$. In practice, good results can be obtained for $\sigma = 1.9999\pi i$, which leads to the row structure $[1, r, 1]$, where $r = -2+9.87\times 10^{-8}$. Assuming solutions of the form $e^{\gamma k}$ leads to $e^{\gamma k}+re^{\gamma(k+1)}+e^{\gamma(k+2)} = 0$, so that $\gamma = -1.11\times 10^{-16}\pm 3.14\times 10^{-4}i$. The solution for a single nonzero RHS term (Fig. 18) does not have compact support, and the end conditions affect the entire system, regardless of n.

Figure 19 shows the spline fit to the step function. The oscillatory behavior in the region of the discontinuity is absent; however, a low-amplitude oscillation (on the order of 10^{-3}) spans the entire curve fit.

Since these splines oscillate in different regions, a "composite spline" (i.e., a cubic spline for regions without discontinuities, and a spline under imaginary tension for the region surrounding a discontinuity) has the potential to confine the remaining oscillations to a specified region.

The two systems can be joined at knot $k = p$ by ensuring that the first and second spatial derivatives are continuous. A cubic spline for $1 \leq k \leq p$ and a spline under tension for $p \leq k \leq n$ lead to the following equations (the equations are similar when the cubic spline and spline under tension reverse positions):

$$2D_1 + D_2 = 3(y_2 - y_1);\ k = 1; \tag{4.6}$$

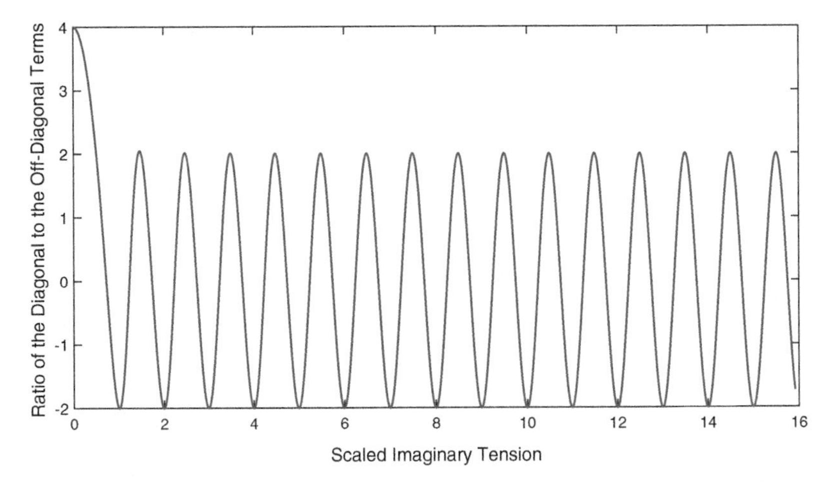

Fig. 17 The ratio of the diagonal term and the off-diagonal term as a function of the scaled imaginary tension $\hat{\sigma} = \text{Im}(\sigma)/2\pi$

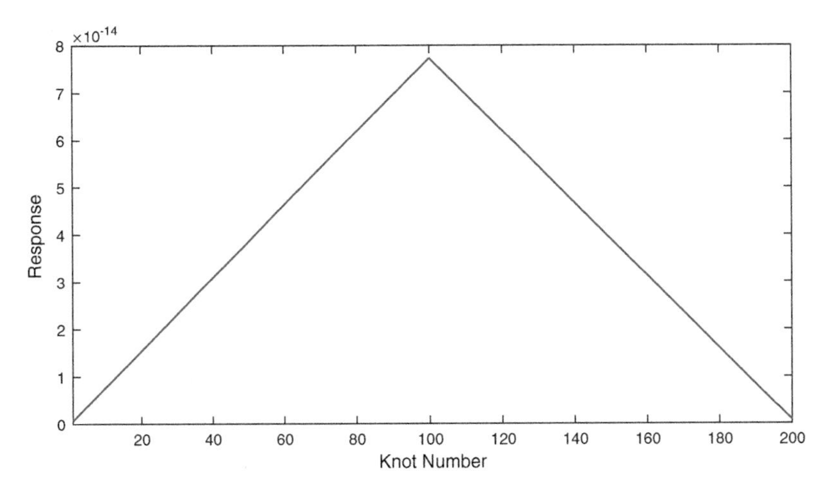

Fig. 18 The response, i.e., $f''(x)$, to a single nonzero RHS term at $k = 100$ for $\sigma = 1.9999\pi i$

$$D_{k+1} + 4D_k + D_{k-1} = 3(y_{k+1} - y_{k-1});\ 2 \leq k \leq p-1; \tag{4.7}$$

$$\left(\frac{1}{\sigma^2 h} - \frac{1}{\sigma \sinh \sigma h}\right) f''(x_{k-1}) + 2\left(\frac{\cosh \sigma h}{\sigma \sinh \sigma h} - \frac{1}{\sigma^2 h}\right) f''(x_k)$$
$$+\left(\frac{1}{\sigma^2 h} - \frac{1}{\sigma \sinh \sigma h}\right) f''(x_{k+1}) = \frac{y_{k+1} - 2y_i + y_{k-1}}{h};\ p+2 \leq k \leq n-1; \tag{4.8}$$

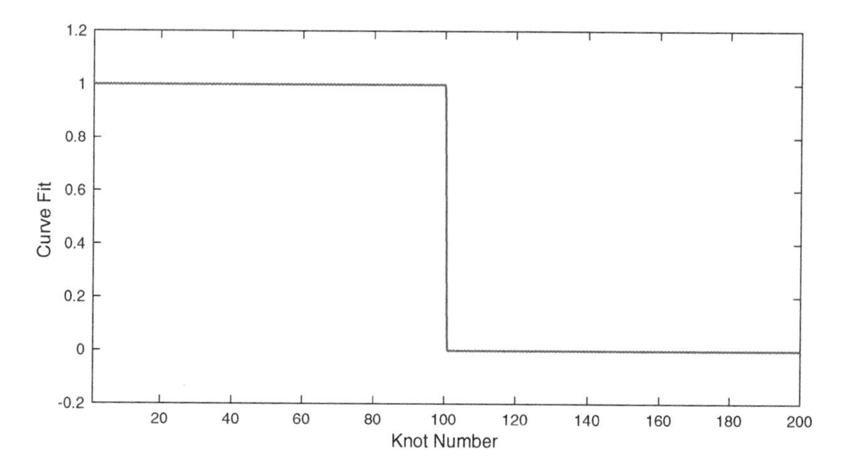

Fig. 19 The spline under tension curve fit for $\sigma = 1.9999\pi i$ for the step function defined by $y_k = 1$ for $1 \leq k \leq 100$ and $y_k = 0$ otherwise. Note that the oscillations in the region of the discontinuity are absent, and an oscillation with an amplitude on the order of 10^{-3} spans the entire curve fit

$$f''(x_n) = 0; k = n; \tag{4.9}$$

$$-f''(x_p) + 4D_p + 2D_{p-1} = 6(y_p - y_{p-1}); k = p; \tag{4.10}$$

$$D_p = f''(x_p)\left(\frac{\cosh \sigma h}{\sigma \sinh \sigma h} - \frac{1}{\sigma^2 h}\right) + f''(x_{p+1})\left(\frac{-1}{\sigma \sinh \sigma h} + \frac{1}{\sigma^2 h}\right) + \left(\frac{y_{p+1} - y_p}{h}\right); k = p. \tag{4.11}$$

Figure 20 shows a composite spline for $p = 100$, with a discontinuity located five knots to the right of the joint. The oscillation amplitude to the left of the discontinuity approaches zero, and to the right, it is on the order of 10^{-4}.

The cubic spline curve fit ends at knot 100, and the spline under tension curve fit begins at knot 101. When the systems are joined, knots 100 and 101 have the same coordinates and data values. Thus, even though both sections contain 100 knots, the entire system spans only 199 knots.

Splines under imaginary tension can be inserted as many times as needed into regions involving discontinuities. Figure 21 shows a three-section composite spline. Here, the oscillations are confined to segments within five knots on either side of the discontinuity. However, in this case, their amplitude has *increased* to approximately 0.017. In Fig. 22, the oscillations are confined to segments within 15 knots on

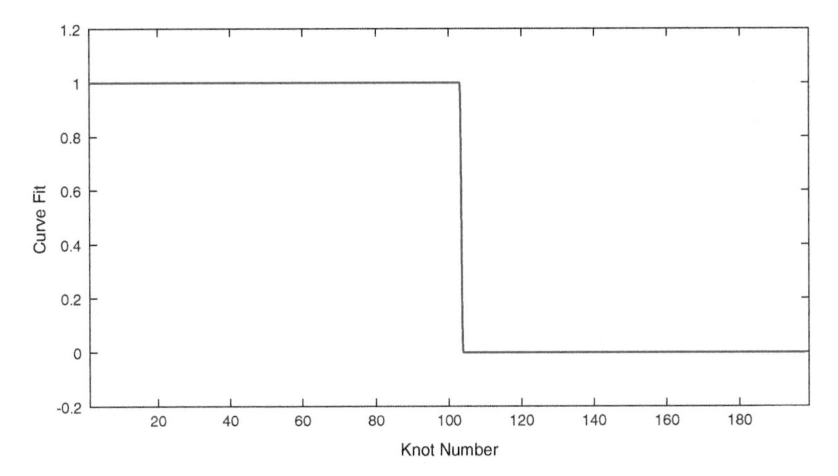

Fig. 20 A "composite spline" consisting of a cubic spine and a spline under imaginary tension joined at $p = 100$. The oscillation amplitude for $k > p$ is on the order of 10^{-4}

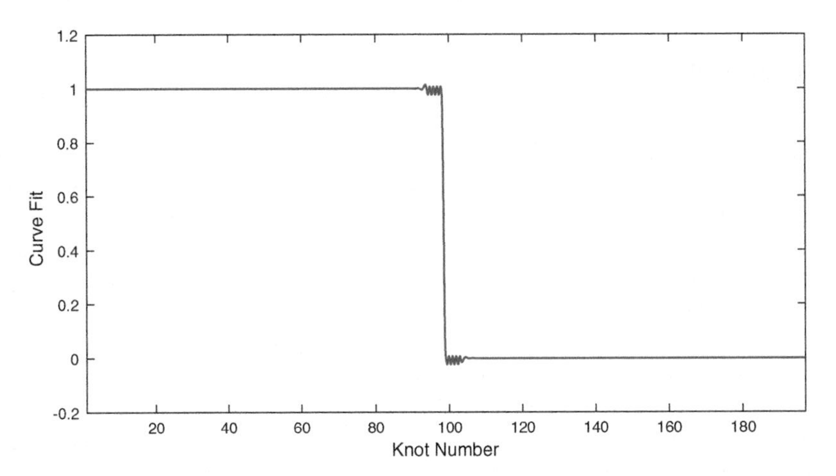

Fig. 21 A "composite spline" consisting of three sections, in which the oscillations are confined to segments within five knots on either side of the discontinuity

either side of the discontinuity, with an amplitude of approximately 0.005, so it appears that the oscillation amplitude for a three-section composite spline may be a linear function of the number of segments in the confined region. The underlying mechanisms that decrease the oscillation amplitude for a composite spline with two sections, but increase the amplitude for three sections, are not understood.

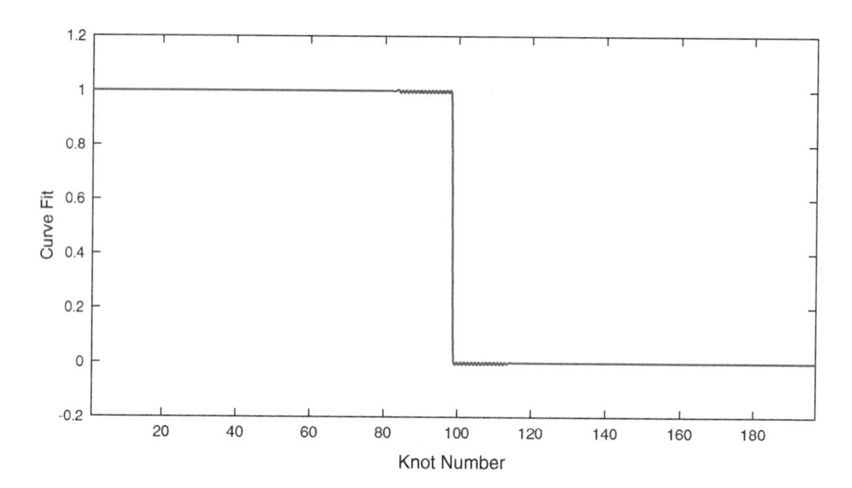

Fig. 22 A "composite spline" consisting of three sections, in which the oscillations are confined to segments within 15 knots on either side of the discontinuity

5 Conclusions

The cubic spline equations form a symmetric tridiagonal Toeplitz system with row structure $[1, 4, 1]$. This generates an oscillatory response for a curve fit of a single nonzero data point. The oscillations cancel for curve fits that do not involve discontinuities. Splines under tension also form a symmetric tridiagonal Toeplitz system characterized by an oscillatory response as a consequence of the row structure $[1, r, 1]$, where $r \geq 4$. Increasing the tension causes the oscillations to decay over a shorter length scale in the region of a discontinuity. Certain values of imaginary tension can lead to a row structure that approaches $[1, -2, 1]$, which removes the oscillations in the region of a discontinuity. However, it also introduces a low-amplitude oscillation that spans the entire curve fit. A composite spline (i.e., a spline under tension in the region surrounding a discontinuity, and a cubic spline elsewhere) can confine the oscillations to an arbitrary region. Further work may be required to fully understand how the confinement process affects their amplitude.

References

1. R.H. Bartels, J.C. Beatty, B.A. Barsky, *An Introduction to Splines for Use in Computer Graphics and Geometric Modeling* (Morgan Kaufmann Publishers, Los Altos, 1987)
2. A.K. Cline, Scalar- and planar-valued curve fitting using splines under tension. Commun. ACM **17**(4), 218–220 (1974)
3. A.A. Ruffa, M.A. Jandron, B. Toni, Parallelized solution of banded linear systems with an introduction to p-adic computation, in *Mathematical Sciences with Multidisciplinary Applications* (Springer, Cham, 2016), pp. 431–464
4. D.G. Schweikert, An interpolation curve using a spline in tension. J. Math. Phys. **45**, 312–317 (1966)

Distributed Membership Games for Planning Sensor Networks

Thomas A. Wettergren and C. Michael Traweek

1 Introduction

Sensor networks are becoming prevalent in a number of monitoring applications [7]. The use of sensor networks for various forms of distributed monitoring of large environments has led to many questions about managing the individual nodes that comprise the system. We concern ourselves with a problem of management of sensor networks with limited central authority. In particular, we develop distributed solutions to the problem of allocating subsets of sensors to individual clusters (or groups) in a distributed sensor field. The subdivision of a sensor field into these individual clusters is beneficial in that it reduces overall sensor field communication requirements. An obvious limiting case is to have each sensor be its own cluster. Such a subdivision is often undesirable in that it causes the sensor field to have excessive false alarms, in that groups of sensors in a cluster can jointly process information to reduce the impact of individual sensors creating a false alarm. On the other hand, to have one single monolithic cluster, while desirable from a false alarm perspective, is usually a communication liability due to the amount of information transfer required between the sensors to maintain a dense network connectivity. In the underwater environment, this network-level communication's capability is especially expensive to maintain, and thus is a critical design consideration. The

---

T. A. Wettergren (✉)
Naval Undersea Warfare Center, Newport, RI, USA
e-mail: thomas.wettergren@navy.mil

C. M. Traweek
Office of Naval Research, Arlington, VA, USA
e-mail: mike.traweek@navy.mil

A. A. Ruffa, B. Toni (eds.), *Advanced Research in Naval Engineering*,
STEAM-H: Science, Technology, Engineering, Agriculture, Mathematics
& Health, https://doi.org/10.1007/978-3-319-95117-1_8

trade-off of these performance characteristics is a goal in sensor network field-level design.

The trade-off to balance objectives through the formation of sensor network cluster complexity (i.e., the determination of which sensors are grouped with which others via communications for joint decision-making) is accomplished as a design optimization problem before sensors are deployed. However, when the environmental characteristics that affect sensor performance are unknown (or highly uncertain), the use of in situ sensor adjustments is critical to developing good system-level performance trade-offs. Whether the adaptation mechanism of the sensors is physical (i.e., moving to a new location to join a different cluster) or purely informational (i.e., communicating with sensors in a different cluster), the mathematical control problem of online determination of cluster membership is the same. The adaptive nature of this online cluster membership determination leads to a group decision process. We develop rules for this group decision process in a manner that does not require negotiation between the agents (sensors) as a group membership game.

Games can be used in two ways to examine this type of sensor network control problem. In particular, a *simulation game engine* can be developed and employed to develop probabilistic representations of expected sensor performance under a variety of actions by both the sensor network and the monitored environment (including the target). That type of game can help to overcome the limit of the impossibly large to physically obtain amount of data required to observe all of the action/consequence permutations between the agents and the target. The second type of game that is useful here is the *group membership game* for the determination of appropriate rules for individual sensors to follow in order to develop a group strategy that leads to the proper allocation of clusters. It is this second type of decision-making game that is the focus of this chapter.

There have been a number of previous studies that show the benefits of groups in sensing [5]. In many sensor networks, groups are employed to improve performance by creating opportunities for data aggregation and improved scalability. We consider groups in the context of a membership game, which is a tool the biological community has developed to examine the rules by which animals join in and separate from groups [4]. In this biological context, it has been shown that the collective behavior of groups (as opposed to individuals) may be beneficial even in search problems [17], which provides a close link to the sensor network problem. Perhaps one of the reasons why group membership games have not been used much is the difficulties in assessing their stability. In particular, in order for a game strategy to be useful in practice, it must guarantee stability. The biological community has observed evolutionarily stable group joining strategies in some biological studies [15].

We use the successes that the biological community has had in observing stable group membership games as an inspiration to develop group membership games for sensor network problems. There have been other studies on game theory for sensor networks [18], yet most of that work has focused on the routing of information between sensors. We instead focus on the allocation of sensors within the network, which is effectively a resource allocation problem (with the resources existing in an information space). There is a significant literature on resource allocation problems with multiple agents [6] and we use those results to guide our focus on group affiliation and group allocation games. While these are effectively coverage games [1], we limit ourselves to a special category of coverage so that we can develop performance guarantees. While others have taken a control systems approach to the analysis of group games in multiagent systems [16], we limit our analysis to games where the strategy space is limited to membership decisions only. Finally, we recognize that others have developed similar games in sensor networks [11], but have not considered the group aspect.

In the sequel, we develop a model for performance of sensor networks that are organized into coordinated clusters (or groups). We then introduce our basic game theory formulation for the strategy space consisting of group membership decisions for individual sensors within the network. Given that framework, we then develop a group affiliation game that can be used to self-assign sensors to groups that are short of ideal capacity. Finally, we develop a group allocation game that has a stable strategy for individual sensors to distribute themselves into groups as a self-organization principle.

2 Sensor Network Model

The type of network model we consider is for a group of N total sensors $\{s_1, \ldots, s_n\}$, that are allocated into G groups, or clusters, of the form $c_g \subset \{s_1, \ldots, s_n\}$. The g-th sensor group, c_g, is a group in that it consists of sensors that share their information through some form of fusion (ranging from simple track-before-detect to complete fusion of raw signature data). Groups may be formed through physical proximity or through directed information sharing. However, in either case the sensor network has the opportunity to evolve through individual sensors making group membership decisions. For combinations due to physical proximity, the group membership decision may be executed through a motion path to relocate the sensor; whereas for other combinations due to directed information sharing, the group membership decision is typically executed in communication networking.

Let the performance of a sensor be given by the sensor search ability, defined as a probability of that sensor detecting any objects of interest that are within a group that the sensor is currently connected to. Specifically, for a sensor s_n this is represented as the following conditional joint probability:

$$\Pr(\text{detect} \mid s_n \in c_g, \text{object} \in c_g) = p_n(c_g, a_n) \tag{1}$$

where a_n is the set of groups c_g for which sensor s_n belongs. This formulation allows an individual sensor to potentially belong to multiple groups and also implies $p_n(c_g, a_n) = 0$ for $c_g \notin a_n$ and $p_n(c_g, a_n) > 0$ for $c_g \in a_n$.

Let $v(c_g)$ be the prior likelihood that the object of interest exists in the region covered by group c_g. Then, the probability of finding an object of interest in a specific group is given by

$$p(c_g) = v(c_g)\left[1 - \prod_{n=1}^{N}(1 - p_n(c_g, a_n))\right]. \tag{2}$$

Assuming that the group detections are independent of one another (i.e., the likelihood of the object being detected in one group has no bearing on the likelihood of the object being detected in any other group), the probability of finding the object of interest can be written as

$$W = \text{Pr(detect)} = \sum_{g=1}^{G} p(c_g) \tag{3}$$

Maximizing the function W within resource constraints is the principal goal of the search or monitoring operation.

We note that for the situation of a single object located somewhere within a domain $\mathscr{D}$ that is completely covered by the groups, we have $\cup_g c_g = \mathscr{D}$ and $\cap_g c_g = 0$, along with $\sum_g v(c_g) = 1$. This special limited situation is the single hidden object in a known domain, known as the Bayesian search problem [10].

3 Game Formulation for Sensor Grouping

The game problems we consider are for games in which each individual sensor decides which group(s) to participate within. In this way, the individual sensors can utilize their own in situ observations of their achievable performance in order to self-organize into a best configuration. If certain game equilibrium conditions are met, then such a game can lead to an optimal configuration of groups for the sensor network. In this context, optimality is guaranteed in a local sense; however, some stronger forms of optimality may be found with limited guarantees. It is important to note that this game formulation allows the network to configure itself in a completely decentralized manner, needing no involvement by a central authority for anything other than setting some contextual problem parameters for each sensor.

In this chapter, we are concerned with games in *strategic form*, whereby the players in the game simultaneously choose their actions (in contrast to games in *extensive form*, whereby players take turns). Such games are appropriate to multiagent systems such as a distributed sensor network, since there is no reason to wait for turn-taking. A strategic game is defined by a tuple of the form $\langle \mathscr{N}, (\mathscr{A}_i)_{i\in\mathscr{N}}, (u_i)_{i\in\mathscr{N}})\rangle$ (following the notation of Bauso [3]). In this definition of a game, the set of *players* is given by $\mathscr{N} = \{1, 2, \ldots, N\}$, the set of *available actions* for player i is given by $\mathscr{A}_i$, and the payoff to player i is given by the *payoff function* u_i. The specific actions taken by player i are given by the action set $a_i \in \mathscr{A}_i$ and generally correspond to the set of decisions made by the player. For a group game, the payoff function for a specific player $u_i = u_i(a_1, a_2, \ldots, a_N)$ depends upon the moves made by all of the players.

The specific form of strategic game under consideration is a *group membership game* [9] in that the decisions/plays made by the agents are to determine their membership (or lack thereof) with respect to the groups involved. As a multiplayer game, this problem has the following components. The players in the game are referred to as *agents* who are involved in a process of making repeated decisions to improve the *system welfare*. This notion of system welfare implies a benefit shared by all when the goal of the group operation is achieved; in this sense, it is a purely non-adversarial game. Yet the game is noncooperative in that the players make decisions independently and without negotiation. There is a *value function* associated with each decision that can be made by the agents. This determines the expected benefit of some decision relative to others independent of the agents. The individual agents each have a perception of their contribution to system welfare (based on their actions), and that is known as their *utility*. There is a *group welfare* which is an achieved benefit that is created by a group of agents who share membership in some subgroup of the whole. We make an assumption that the individual agents within a group can all perceive the group welfare of that particular group.

We consider a group $\mathscr{N}$ of N players who are representative of the physical sensors $\{s_1, s_2, \ldots, s_N\}$ that are available for the group membership game. Let the action set a_n be the set of sensor groups c_g that sensor s_n participates within. This action set corresponds to the moves in the game that are taken by the sensor, where a move corresponds to joining a specific sensor group. The game's value function is the search object's prior likelihood $v(c_g)$ of being in the domain of group c_g. Note that this selection of value function prescribes a natural sense of value to specific groups, whereby groups with higher value are those in which it is more likely to find the object of interest.

The group welfare of an individual group is the probability that the search effort applied by that group is successful in the goal of finding the hidden object. Thus, the group welfare is given by the probability $p(c_g)$ in Eq. (2). The total system welfare is the total probability of detecting the hidden object, given by the summation of the group welfares over all of the groups, as in Eq. (3). Thus, the payoff u_n for an individual player (agent) can be described by the perceived contribution of the individual to the overall group welfare. In that context, the individual utility that is perceived by agent s_n is given by

$$u_n(A) = \sum_{c_g \in a_n} \frac{p(c_g)}{M_g} \tag{4}$$

where $M_g = \sum_n I_{a_n}(c_g)$ is the number of agents in group c_g, and $A = \{a_1, \ldots, a_N\}$ is the set of actions of all of the agents. This utility (or payoff) for an individual has the benefit that it becomes the group welfare of Eq. (2) when individuals are restricted to a single group and the individual utilities are summed over all agents in that group. Similarly, the utility leads to the system welfare of Eq. (3) when it is applied to all of the agents and all individual utilities are summed. Thus, it is an appropriate payoff function for a game model of systems that consider either the group or system welfare as a performance metric.

4 The Group Affiliation Game

The first game problem we consider is one where the sensors decide to affiliate with groups (or not) based on a local benefit of staying where they are or joining the group. This type of strategy allows for robust changes to sizes of groups that are near to one another, but as a practical matter it presumes that sensors are already deployed with a notion of overall group composition. Thus, we assume that there exists a set $\mathcal{N}$ of N sensors $\{s_1, \ldots, s_N\}$ that have been deployed into G groups $\{c_1, \ldots, c_G\}$ according to some a priori information. We assume that this initial deployment was made in a meaningful way, such that it is near to the optimal initial separation into groups.

As the network evolves over time, there are often opportunities for sensors to reconsider their particular membership due to changes in overall sensor network composition, changes in the environmental performance of sensor nodes, or changes in the network's goals. For example, as time evolves and the overall field is potentially degraded, the addition of new sensors provides an opportunity for improvement, and allowing the individual sensors to make their own decisions for group membership allows for appropriate sizing of groups (since a sensor will be less likely to join a group that is already overpopulated). Furthermore, as the environment changes over time, there may be opportunities to improve performance by a sensor changing its membership from one group that is now overpopulated to another that is underpopulated. This is a game with an evolutionary strategy for selecting membership that we refer to as *the group affiliation game*. As the game evolves with individual agents making group membership decisions, we hope to achieve a convergence to a fixed overall group topology. This can be shown if *evolutionary stability* holds for the strategy space of the game.

We examine a simplified version of this game, where individual agents make decisions to join or not join the existing groups based upon the benefit of the agent in the group relative to the agent staying alone (i.e., a group of itself). In this way, we develop a rule for maximal reasonable group sizes based upon the relative benefits of differing group values. The overall game can be decomposed into a set of repeated plays of this very simple decision over all of the agents. This game ignores the concept of individual agent utility (as given by Eq. (4)) and instead focuses on simple assessments of group performance with and without the agent in question.

Let us assume that there is an existing group c_g that contains N_g agents. Furthermore, assume that an individual agent is not in a group (or more specifically, it is in a group of size one). Let that agent's individual group be denoted as c_0. We model an arbitrary agent's performance in group c_g by:

$$p_g = p_n(c_g, a_n) \qquad \forall \{s_n : c_g \in a_n\} \tag{5}$$

The model of Eq. (5) implies that all agents are identical and anonymous. Furthermore, assume that an arbitrary agent's performance in group c_0 is given by:

$$p_0 = p_n(c_0, a_n) \qquad \forall \{s_n : c_0 \in a_n\} \tag{6}$$

For c_0 to be an individual agent group, we require the additional implied constraint $\{c_0 \in a_n \Rightarrow c_0 \notin a_m \ \forall m \neq n\}$. We consider the added benefit that is achieved by the agent in question when making the choice to join either c_g or c_0.

If the agent decides to join group c_g, the group welfare (as given by Eq. (2)) for group c_g is increased by an amount given by:

$$\begin{aligned}\Delta p(c_g) &= v(c_g)\left[1 - (1 - p_g)^{g+1}\right] - v(c_g)\left[1 - (1 - p_g)^{g}\right] \\ &= v(c_g)p_g(1 - p_g)^{g}\end{aligned} \tag{7}$$

which is equivalent to the change in the total system welfare for such a decision. However, if the agent decides to "join" its own group of c_0, the group welfare for group c_0 is increased (increased relative to not having the agent there) by an amount given by:

$$\Delta p(c_0) = v(c_0)\left[1 - (1 - p_0)\right] - 0 = v(c_0)p_0 \tag{8}$$

which is again equivalent to the change in the total system welfare for such a decision.

We view this simple decision as a repeated two-player game (where the two players are the group and the joining individual) and seek to find if an evolutionary stable joining strategy exists. Let the payoff function u_i for the agent in question be given by the change to group welfare that is created by the agent's decision on group membership. In the two-player game strategy space, there are four potential outcomes based on whether the individual stays alone (action 0) or tries to join the group (action 1) and whether the group desires the individual (action 1) or refuses the individual (action 0). For such a game, the strategy space consists of four outcomes $J(0, 0)$, $J(0, 1)$, $J(1, 0)$, and $J(1, 1)$, where $J(i, j)$ is for the individual performing the strategy with action i and the group performing the strategy with action j. Since we only consider membership when both the group and individual select it, we have $J(1, 1) = \Delta p(c_g)$ and $J(0, 0) = J(0, 1) = J(1, 0) = \Delta p(c_0)$.

It is clear from this simple game that the pure strategy of the individual joining decision (action 1 for the individual) is an evolutionary stable strategy [20] when the following condition holds true:

$$J(1,1) > J(0,1) \tag{9}$$

Furthermore, we note that the corresponding pure strategy for the group is evolutionarily stable when a similar condition ($J(1,1) > J(1,0)$) holds true. Thus, group membership (which requires both the individual and group to choose a joining decision) is a stable strategy when the condition $\Delta p(c_g) > \Delta p(c_0)$ holds true, which is given by

$$g < g^* = \frac{\log(p_0/p_g) + \log(v(c_g)/v(c_0))}{\log(1 - p_g)} \tag{10}$$

This requirement provides a stable maximum size for a given group, beyond which individuals should choose other alternatives. Numerical examples of this decision-theoretic group membership rule are given by the curves shown in Figs. 1 and 2. These curves (and the associated calculation of g^* as per Eq. (10)) can be utilized by groups and individuals as a rule for deciding when to connect to a group of sensors during periods of replenishment or reconfiguration. In this way, the game provides an opportunity for the system to adapt without supervisory control. We

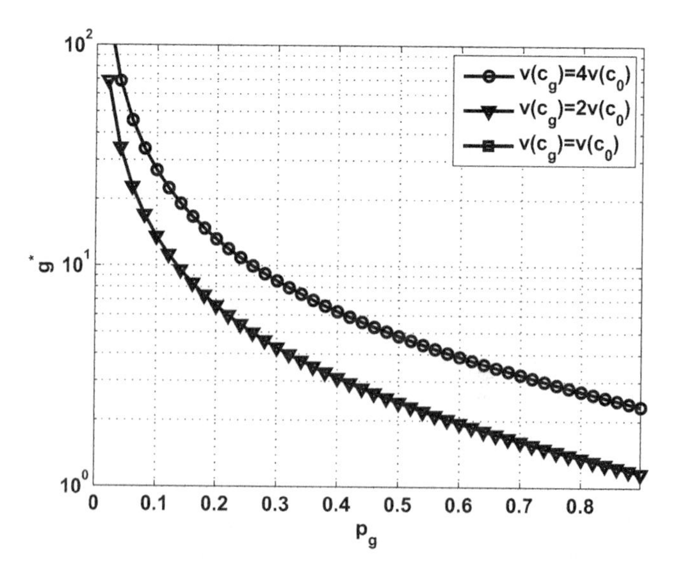

Fig. 1 Maximum desirable size g^* for a sensor group based on Eq. (10) from the group affiliation game. This case is for values where the sensor detection performance is the same in the group as on its own ($p_0 = p_g$). Note that there is no benefit to being in the group when $v(c_g) = v(c_0)$ for this case

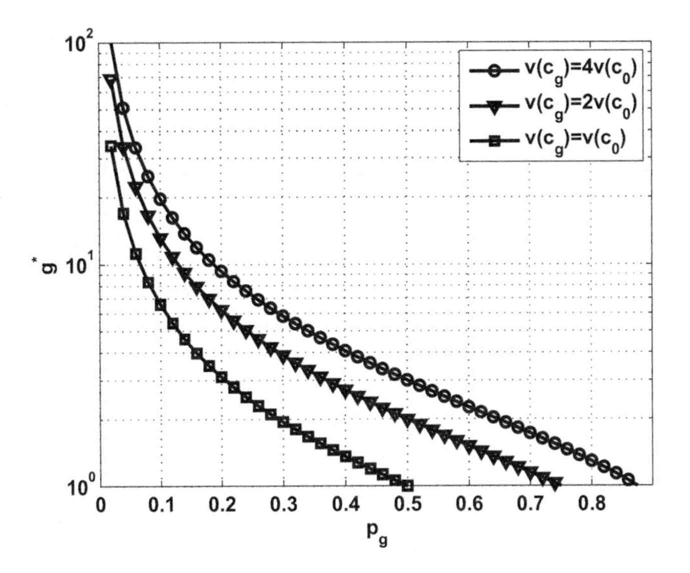

Fig. 2 Maximum desirable size g^* for a sensor group based on Eq. (10) from the group affiliation game. This case is for values where the sensor detection performance is better in the group than on its own ($p_0 = p_g/2$)

note that our affiliation game framework is similar to that in [2], but distinct in that their formulation uses a *free entry game* where individuals can join groups without concern of the group's desire for membership.

While this simple repeated membership game analysis is useful for determining optimal sizes of groups, it has two limitations: (1) the decisions are limited to a trivial alternative when not joining the group, and (2) the decisions have no reliance on the overall number of agents. To avoid these problems, we next consider the analysis of the game with the individual agent utility of Eq. (4).

5 The Group Allocation Game

While the group affiliation game provides guidance on when a sensor should affiliate itself with a particular group, it does not consider the overall size of the network. Thus, it is most appropriate to be used only as an adaptation rule for an already functioning network. When applied as a strategy for a sensor network to self-organize, the affiliation game will lead to either a few well-filled groups and other nearly empty groups (in the case when there are small numbers of available sensors) or cycling of sensor between groups as all groups reach their g^*-level and "extra" sensors cannot find a place to fit. It is clear that a different game that accounts for the overall number of available sensors is required if a game formulation is to be used for initial self-organization of network groups. The game that we develop for such a problem is *the group allocation game*.

Whereas the group affiliation game uses principles of behavioral ecology games, the group allocation game uses principles of economic games. In particular, we use a particular type of *distributed welfare game* [12] to divide a shared set of available resources among individual players in a manner that distributes the welfare so as to achieve improved overall system performance. The distributed welfare game that we consider defines the resources to be shared as the groups $\{c_1, \ldots, c_G\}$ that individual sensors can be members of. The players of the game are the sensors $\{s_1, \ldots, s_N\}$, and the welfare to be distributed is the overall system welfare W, given by the joint probability of detection in Eq. (3).

Within distributed welfare games, an important subcategory of game is a *social utility system*. In this type of game, the players behave noncooperatively in that they do not directly try to maximize the overall system welfare but rather are only trying to maximize their own utility. While that may not seem to benefit the whole, the particular feature of a social utility system is that it has an individual utility that is measured in such a way so that both a Nash equilibrium for the system exists and that equilibrium has a value of system welfare that is at least one-half of the welfare of the optimum. The first of these conclusions implies that all players only need to follow a strategy to maximize their own utility and that is all that is required for the system to achieve equilibrium. The second conclusion implies that the equilibrium that is achieved will be within one-half of the value of overall system welfare that would be achieved with an optimal solution from a complete centralized control of the entire system. Thus, the social utility system is an important category of distributed welfare game for potential use with distributed sensing problems, as it reduces the need for central oversight.

In order for a distributed welfare game to be a social utility system, there are three conditions that must be met [19]. The first condition is that the system welfare must be a submodular function with respect to the players. This condition provides a diminishing return for the addition of more players to the game. The second condition is that the utility of an individual player must be greater than or equal to the change in social welfare that occurs if the player was removed from the game. This condition is known as the *Vickery utility* in economics, and it provides a lower bound on individual utility that is similar to that found in auction systems. The third condition is that the sum of the individual player utilities must be no greater than the total system welfare. This condition guarantees that individual utility is reflective of a benefit toward total group reward.

Consider a sensor network with sensors that can only belong to one group c_g at a time. We now demonstrate that a distributed welfare game consisting of such sensors that follow the utility function of Eq. (4) with the system welfare of Eq. (3) is a social utility system. We first consider the condition of submodularity of the system welfare function.

The sensor joint probability of detection of Eq. (3) is submodular for agents that are identical and anonymous, as shown by the following. The welfare function $W(S)$ for a set of players S is submodular if for any player sets $S_0 \subseteq S_1 \subseteq S$, we have

$$W(S_0 \cup s_n) - W(S_0) \geq W(S_1 \cup s_n) - W(S_1) \tag{11}$$

for any $s_n \in S$. Without loss of generality, let $p_g = p_n(c_g, s_n)$ for any sensor s_n in group c_g (via the identical and anonymous assumption), and let S_0 contain N_0 sensors and S_1 contain $N_1 \geq N_0$ sensors. Recognizing that all welfare is a simple sum of welfare over each group, we analyze the relation for an arbitrary group g. In particular, for a given group g we have for individual group welfare W_g the following:

$$\begin{aligned} W_g(S_0 \cup s_n) - W_g(S_0) &= v_g \left[1 - \prod_{n=1}^{N_0+1} (1 - p_g) \right] - v_g \left[1 - \prod_{n=1}^{N_0} (1 - p_g) \right] \\ &= v_g \left[(1 - p_g)^{N_0} - (1 - p_g)^{N_0+1} \right] \\ &= v_g p_g (1 - p_g)^{N_0} \end{aligned} \tag{12}$$

and in a similar way, we have

$$W_g(S_1 \cup s_n) - W_g(S_1) = v_g p_g (1 - p_g)^{N_1} \tag{13}$$

where we have used $v_g = v(c_g)$ for ease of exposition. Now, it is simple to show

$$\begin{aligned} \Delta W_g &\equiv \left[W_g(S_0 \cup s_n) - W_g(S_0) \right] - \left[W_g(S_1 \cup s_n) - W_g(S_1) \right] \\ &= v_g p_g (1 - p_g)^{N_0} \left[1 - (1 - p_g)^{N_1 - N_0} \right] \\ &\geq 0 \end{aligned} \tag{14}$$

for $N_1 \geq N_0$, where the inequality is guaranteed since every term in the product is greater than or equal to zero. Since $\Delta W_g \geq 0$ for all groups g, and the total system welfare is the simple sum of individual group welfares W_g, we have that

$$\Delta W = \Delta \sum_{g=1}^{G} W_g = \sum_{g=1}^{G} \Delta W_g \geq 0 \tag{15}$$

which implies $W(S_0 \cup s_n) - W(S_0) \geq W(S_1 \cup s_n) - W(S_1)$. Thus, the submodularity condition of Eq. (11) holds for the sensor group membership game system welfare function.

The second and the third condition for social utility we consider concurrently. In particular, we note that for this system (with sensors restricted to a single group at a time) the utility formulation of Eq. (4) is equivalent to distributing the share of system welfare equally to each individual in the group. To see this, we consider a group c_g with N_g sensors. For such a group, we have the sum of the individual utilities given by

$$\sum_{n=1}^{N_g} u_n = \sum_{n=1}^{N_g} \frac{p(c_g)}{N_g} = p(c_g) \sum_{n=1}^{N_g} \frac{1}{N_g} = p(c_g) \tag{16}$$

which shows that the sum of utilities that is given by this form of utility is equivalent to the system welfare (and thus is no greater than the system welfare), demonstrating the third condition. The second condition for this case is a consequence of the third in that the change in system welfare for the removal of each individual must be equal to their utility if the utilities are the same and their sum equals the total system welfare. Since this condition holds within a group and the individual's effect on system welfare is limited to only that group, it also holds for the entire system. Thus, all conditions are met and a system of sensors restricted to single groups with utility of Eq. (4) and system welfare of Eq. (3) is a social utility system. As such, we can be guaranteed that a system that follows the game's rules will obtain a Nash equilibrium in an evolutionarily stable manner, and furthermore, that equilibrium solution will have a system welfare within a guaranteed bound of the optimal solution that could be obtained through optimization of centralized control.

We next demonstrate how a system of sensors can make group membership decisions in order to maximize their utility of Eq. (4). In particular, these sensors need to maximize their utility based upon their assumption about other sensors. One popular way to do this is to use the notion of *fictitious play* [14]. In a group game employing fictitious play, an individual player informs themselves of the potential actions of the other players by an examination of their past actions. In particular, each player uses the empirical distribution of other players' actions as a model of their strategy for the game. Obviously, such a game requires detailed examination of the behavior of others which may not be always available (i.e., we may be able to observe others staying in, joining, or leaving the group we are in, but not other groups). In such cases, extensions to fictitious play such as joint strategy fictitious play [8] and joint strategy fictitious play with inertia [13] can be used to create approximate representations of the other players' strategies in an efficient manner.

To illustrate the performance of the social utility system for the group allocation game, we performed numerical simulation experiments for a group of sensors self-organizing into appropriate sensing groups. We consider a set of $N = 12$ identical and anonymous sensors that are to be allocated within $G = 4$ groups for conducting distributed monitoring in a domain $\mathscr{D} \subset \mathbb{R}^2$. Let $|c_g|$ be the cardinality associated with group c_g. We assume that the groups have been predefined due to environmental considerations within the domain, and the prior likelihood of finding the object(s) of interest within each group is given by $\{v(1), v(2), v(3), v(4)\} = \{0.4, 0.4, 0.1, 0.1\}$. Thus, we have two groups (c_1 and c_2) which are likely to contain our object of interest and two groups (c_3 and c_4) which are much less likely but may still contain the object. Note that $\sum_g v(c_g) = 1$ so this example meets the special conditions of the Bayesian search problem as described in Sect. 2. We also assume that the probability of detecting an object in a given group, when the sensor is in the group and the object is in the group, is given (as in Eq. (1)) by $\{p_n(1, 1), p_n(2, 2), p_n(3, 3), p_n(4, 4)\} = \{0.8, 0.4, 0.8, 0.8\}$. We assume that sensors can only belong to one group and can only detect the object that is in that group, so $p_n(i, j) = 0 \; \forall i \neq j$, and since all sensors are identical and anonymous, $p_n(i, i) = p_m(i, i) \; \forall n, m$. Note that the detection probabilities show that one of the groups (group c_2) has a detection performance that is significantly degraded from the other groups, which is representative of environmental obstructions and/or limited clarity in view.

Simulations were run with the $N = 12$ sensors all starting in group c_1 (such that $|c_1| = 12$, and $|c_i| = 0$ for $i = 2, 3, 4$), and executed a simulation of 100 steps of a group decision process. At each step, each sensor (chosen in random order) individually considers the group that it is in and makes a decision to either stay in that group or join another group. The sensor makes this decision by maximizing its expected change in utility, with utility given by Eq. (4). The expected performance of other sensors is given by a frequency of occurrence of joining/staying of past performance of each sensor in the group. Based on this, the sensor decides which group to join (or to stay in the current group) for the next iteration. We also add in an error with probability p_ε that represents an individual sensor probability of not observing other groups. When that occurs, the given sensor is forced to stay in its current group regardless of potential utility. This can account for a number of intermittent errors that may occur in practice. For these simulations, we use $p_\varepsilon = 0.05$ for a 5% chance of error in any given sensor decision.

In Fig. 3, we illustrate the results of one of these simulation runs. In the figure, we show 12 separate plots, one for each of the 12 sensors. Note that the sensors make intermittent changes to their group membership and that other sensors respond later to those changes. At the end of the 100 iterations, the sensors have reached a configuration with the following memberships:

$$|c_1| = 5, \;\; |c_2| = 5, \;\; |c_3| = 1, \;\; |c_4| = 1.$$

This configuration is consistent with the first two groups having much higher prior likelihoods $v(c_g)$ than the latter two groups. Thus, the group allocation game was successful in creating a self-organized distribution of group memberships with no central oversight. In Fig. 4, we plot the resulting system welfare W (given by Eq. (3)) for this simulation run. Note that even though the individual sensors demonstrate intermittent switching between groups, the overall performance in the system welfare stays relatively near the optimum after a very brief initial increase. Also, note that the allocation at the end is not the true system optimum, as there are some group combinations (such as the one near time step 83) that give slightly better performance. However, we note that this only illustrates that this particular

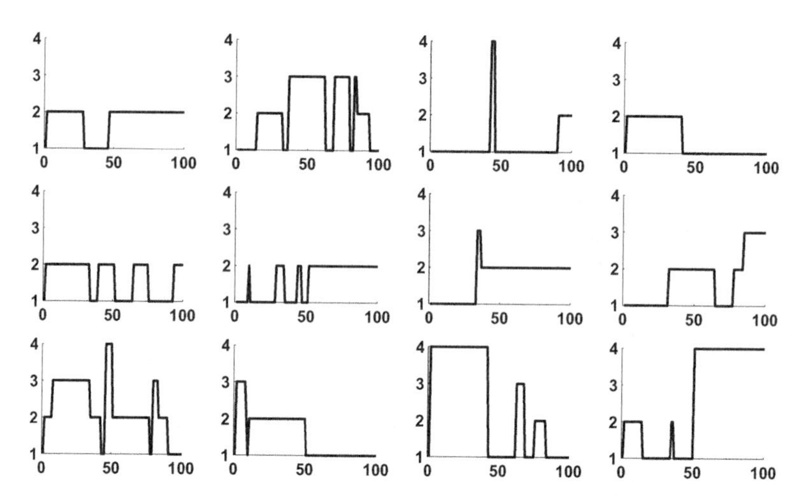

Fig. 3 Group memberships for each of the 12 simulated sensors over 100 time steps. The vertical axis of each plot represents that group number for that sensor at each time step

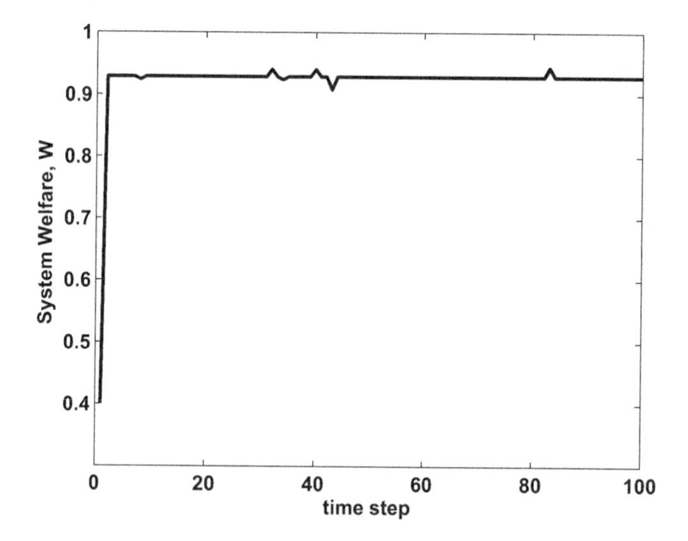

Fig. 4 The total system welfare as given by Eq. (3) for the simulation run with group memberships as shown in Fig. 3

system has multiple Nash equilibria, and the evolutionary joining strategy is only guaranteed to reach one of them, not a particular one.

In Fig. 5, we illustrate the results of another simulation run. In this simulation, the sensors all start in group c_1 as in the previous simulation; however, at the end of iteration 50, we remove two of the sensors from the simulation. This corresponds to a malfunction and/or battery drainage for those sensors. One of the expected benefits of the self-organization property of the group allocation game is to adapt to

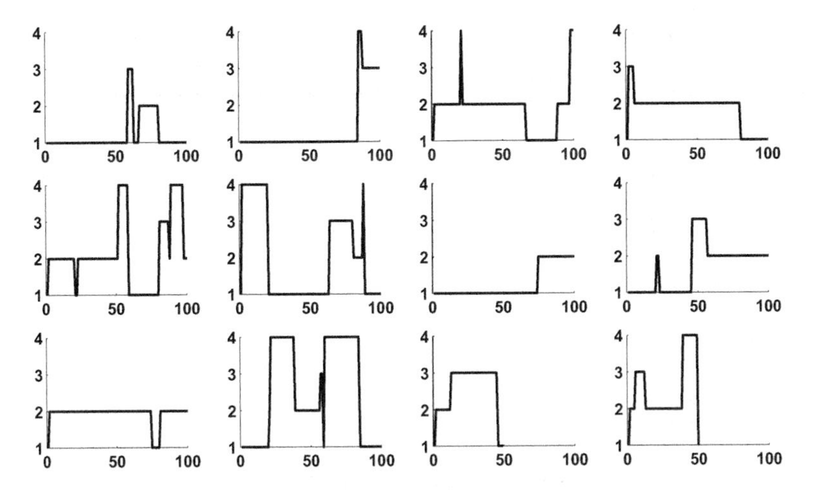

Fig. 5 Group memberships for each of the 12 simulated sensors over 100 time steps, with a removal of the last 2 sensors (two rightmost plots on bottom row) after 50 time steps. The vertical axis of each plot represents the group number for that sensor at each time step

unexpected changes in the availability of players. At the end of the 100 iterations, the sensors have reached a configuration with the following memberships:

$$|c_1| = 4, \ |c_2| = 4, \ |c_3| = 1, \ |c_4| = 1.$$

While being a modification from the configuration with the full 12 sensors, this 10-sensor configuration is still consistent with the first two groups having much higher prior likelihoods $v(c_g)$ than the latter two groups. Thus, the group allocation game was successful in adapting the memberships in a self-organized manner with no central oversight. In Fig. 6, we plot the resulting system welfare W (given by Eq. (3)) for this simulation run. In this plot, the immediate performance degradation due to the loss of two sensors at iteration 50 is clear. However, the system is able to quickly adapt and the sensors regroup into modified memberships to regain much of the lost performance and achieve a stable equilibrium with the new number of available sensors.

6 Conclusion

We have developed two forms of group membership games for sensor networks comprised of a fixed number of groups. In the group affiliation game, we use game theory to develop rules for joining based on a maximal perceived group size. In the group allocation game, we form a social utility system of distributed welfare games to provide rules for joining that lead to self-organization. Both of these games

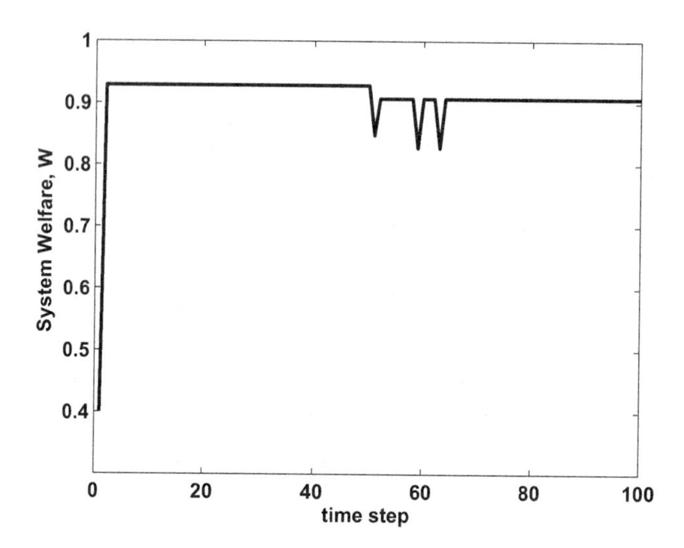

Fig. 6 The total system welfare as given by Eq. (3) for the simulation run with group memberships as shown in Fig. 5

illustrate the ability of game theory to be applied to group membership such that sensor network systems can perform optimal adaptation with no central oversight. Furthermore, these particular games require no negotiation between sensors and are therefore applicable in situations with limited communications between sensors.

References

1. X. Ai, V. Srinivasan, C.-K. Tham, Optimality and complexity of pure Nash equilibria in the coverage game. IEEE J. Sel. Areas Commun. **26**(7), 1170–1182 (2008)
2. B.W. Andrews, K.M. Passino, T.A. Waite, Social foraging theory for robust multiagent system design. IEEE Trans. Autom. Sci. Eng. **4**(1), 79–86 (2007)
3. D. Bauso, *Game Theory with Engineering Applications* (SIAM, Philadelphia, 2016)
4. D. Boyer, O. Miramontes, G. Ramos-Fernández, J.L. Mateos, G. Cocho, Modeling the searching behavior of social monkeys. Phys. A **342**, 329–335 (2004)
5. O. Boyinbode, H. Le, A. Mbogho, M. Takizawa, R. Poliah, A survey on clustering algorithms for wireless sensor networks, in *Proceedings of 13th International Conference on Network-Based Information Systems* (2010), pp. 358–364
6. Y. Chevaleyre, P.E. Dunne, U. Endriss, J. Lang, M. Lemaitre, N. Maudet, J. Padget, S. Phelps, J.A. Rodriguez-Aguilar, P. Sousa, Issues in multiagent resource allocation. Informatica **30**, 3–31 (2006)
7. D. Culler, D. Estrin, M. Srivastava, Overview of sensor networks. IEEE Comput. Mag. **37**(8), 41–49 (2004)
8. D. Fudenberg, D. Kreps, Learning mixed equilibria. Games Econ. Behav. **5**(3), 320–367 (1993)
9. L.A. Giraldeau, T. Caraco, *Social Foraging Theory* (Princeton University Press, Princeton, 2000)

10. B.O. Koopman, *Search and Screening: General Principles with Historical Applications* (Pergamon Press, New York, 1980)
11. J.R. Marden, A. Wierman, Distributed welfare games with applications to sensor coverage, in *Proceedings of 47th IEEE Conference on Decision and Control* (2008), pp. 1708–1713
12. J.R. Marden, A. Wierman, Distributed welfare games. Oper. Res. **61**(1), 155–168 (2013)
13. J.R. Marden, G. Arslan, J.S. Shamma, Joint strategy fictitious play with inertia for potential games. IEEE Trans. Autom. Control **54**(2), 208–220 (2009)
14. D. Monderer, L.S. Shapley, Fictitious play property for games with identical interests. J. Econ. Theory **68**, 258–265 (1996)
15. G.D. Ruxton, C. Fraser, M. Broom, An evolutionary stable joining policy for group foragers. Behav. Ecol. **16**(5), 856–864 (2005)
16. E. Samsar-Kazerooni, K. Khorasani, Multi-agent team coordination: a game theory approach. Automatica **45**, 2205–2213 (2009)
17. M.C. Santos, E.P. Raposo, G.M. Viswanathan, M.G.E. da Luz, Can collective searches profit from Lévy walk strategies. J. Phys. A: Math. Theor. **43** (2009). Article number 434017
18. H.Y. Shi, W.L. Wang, N.M. Kwok, S.Y. Chen, Game theory for wireless sensor networks: a survey. Sensors **12**, 9055–9097 (2012)
19. A. Vetta, Nash equilibria in competitive societies, with applications to facility location, traffic routing and auctions, in *Proceedings of 43rd Annual IEEE Symposium on Foundations of Computer Science* (2002), pp. 416–425
20. J.W. Weibull, *Evolutionary Game Theory* (MIT Press, Cambridge, 1998)

Statistical Models of Inertial Sensors and Integral Error Bounds

Richard J. Vaccaro and Ahmed S. Zaki

1 Introduction

Inertial sensors such as gyroscopes and accelerometers are important components of inertial measurement units (IMUs). Inertial sensors provide an output that is proportional to angular velocity or linear acceleration. The output signal is corrupted by additive noise plus a random drift component. This drift component, also called bias, is modeled using different types of random processes. This chapter considers the random components that are useful for modeling modern tactical-graded MEMS sensors. Other important issues related to sensor performance such as scale-factor errors, quantization effects, and temperature effects are not considered in this chapter. See [1] for a description of these effects. The additive noise and random drift components that corrupt a sensor signal contribute to errors in the first and second integrals of the sensor output. The purpose of this chapter is to derive formulas for bounding these errors. The chapter is organized as follows.

Section 2 introduces the mathematical models describing the random components of a sensor signal. These components include additive white noise, integrated white noise (random walk), and integrated filtered white noise (called Gauss–Markov components). All of these components are needed to model modern MEMS

R. J. Vaccaro
Department of Electrical, Computer, and Biomedical Engineering, University of Rhode Island, Kingston, RI, USA
e-mail: vaccaro@ele.uri.edu

A. S. Zaki (✉)
Naval Undersea Warfare Center, Division Newport, Newport, RI, USA
e-mail: ahmed.zaki@navy.mil

A. A. Ruffa, B. Toni (eds.), *Advanced Research in Naval Engineering*, STEAM-H: Science, Technology, Engineering, Agriculture, Mathematics & Health, https://doi.org/10.1007/978-3-319-95117-1_9

sensors. This section also introduces Allan variance, a well-known statistic for characterizing the behavior of inertial sensors.

In Sect. 3, formulas are derived for bounding the errors in the first and second integrals of a sensor output. These errors are caused by the random noise and drift components in the sensor signal. The error bound on the first integral is shown to be a simple function of the sensor's Allan variance, which may be estimated from a sensor calibration signal (sensor output in the absence of motion). The error bound on the second integral is obtained by numerically integrating the first-integral bound. In order to perform the numerical integration, the integrand, which is a function of the Allan variance, must be evaluated at a dense grid of points. Given a sensor calibration signal, the Allan variance is not easily estimated on a dense grid of points. One way to proceed is to estimate the parameters of a mathematical model for a sensor's random components, and use this model to evaluate the sensor's Allan variance on a dense grid of points.

Section 4 deals with estimating the parameters of a sensor model by least-squares and weighted least-squares curve fitting to a set of calculated Allan variance points. Section 5 provides a summary and describes future work. An Appendix provides MATLAB programs for simulating a calibration signal, for calculating Allan variance, and for estimating sensor parameters.

2 Mathematical Models and Allan Variance

The output signal, $y(t)$, of a sensor in the absence of motion will be referred to as a *calibration signal*, and can be described by the following equation:

$$y(t) = b(t) + n(t) \tag{1}$$

where $b(t)$ is the bias (random drift), and $n(t)$ is observation noise, which is assumed to be white noise with spectral density R. The bias term will be modeled with one or more components. The first bias component, $b_0(t)$, is a random process consisting of the integral of a white-noise process $w_0(t)$ with spectral density Q_0. The integral creates a nonstationary random process whose variance grows with time, which is a good model for the drift of a sensor. This bias component is modeled as follows:

$$\dot{b}_0(t) = w_0(t), \quad E[w_0(t_1)w_0(t_2)] = Q_0\delta(t_1 - t_2). \tag{2}$$

It has been observed that, prior to the year 2000, many tactical grade MEMS sensors were well-modeled by (1) and (2), with $b(t) = b_0(t)$ [2–4]. However, improvements in MEMS technology have resulted in the need to use additional components to adequately model sensor drift. These additional components are so-called Gauss–Markov (GM) random processes. The ith GM component consists of white noise with spectral density Q_i passed through a single-pole filter with stable pole located at $-\alpha_i$, and is characterized by the parameters (α_i, Q_i). The mathematical model for the ith GM component is

Table 1 Typical units for spectral densities Q_i, $i = 0, 1, \cdots, C$ and R for rate gyros and accelerometers

Spectral density	Rate gyro	Accelerometer
R	deg^2/h	m^2/h^3
Q_i	deg^2/h^3	m^2/h^5

$$\dot{b}_i(t) = -\alpha_i b_i(t) + w_i(t), \quad E[w_i(t_1)w_i(t_2)] = Q_i\delta(t_1 - t_2). \tag{3}$$

For rate gyros, the additive noise component is referred to as the *angle random walk* (ARW) and the bias component $b_0(t)$ is referred to as the *rate random walk* (RRW). The complete model for a sensor calibration signal is given by (1) with the bias term given by combining (2) and (3) to obtain

$$b(t) = b_0(t) + \sum_{i=1}^{C} b_i(t). \tag{4}$$

All of the white-noise processes, $n(t)$ and $w_i(t)$, $i = 0, \cdots, C$, are assumed to be zero mean and statistically independent. The parameters that are needed to describe the statistical behavior of a sensor calibration signal consisting of additive noise and multiple drift components are the following: $[R, Q_0, (\alpha_1, Q_1), \cdots (\alpha_C, Q_C)]$. Typical units for Q_i and R are shown in Table 1.

2.1 Allan Variance

Allan variance, denoted $a(\tau)$, is computed for different values of a smoothing parameter, τ [1, 5]. In order to compute the Allan variance associated with a record of calibration data, the signal is averaged over length-τ intervals. The resulting average values are denoted $\bar{y}_k$, where

$$\bar{y}_k = \frac{1}{\tau}\int_{(k-1)\tau}^{k\tau} y(\lambda)d\lambda. \tag{5}$$

The definition of Allan variance is [1]

$$a(\tau) = \frac{1}{2}E[(\bar{y}_{k+1} - \bar{y}_k)^2] \tag{6}$$

where $E[\cdot]$ denotes expected value. Note that the quantity that is being squared in this equation is a zero-mean random variable, and so the expected value is the variance of $(\bar{y}_{k+1} - \bar{y}_k)$. In the previous two equations, k is an arbitrary positive integer. The definition of Allan variance is independent of k because the expected value in (6) is the same for any value of k. Thus, given a finite amount of calibration data obtained over an interval of time of length D, the following estimator for the variance in (6) may be used to obtain an estimate of the Allan variance:

$$\hat{a}(\tau) = \frac{1}{2(N_\tau - 1)} \sum_{k=1}^{N_\tau - 1} (\bar{y}_{k+1} - \bar{y}_k)^2, \tag{7}$$

where N_τ is the largest integer that is smaller than D/τ.

2.2 *Single Drift Component*

Consider the case in which the sensor calibration signal is modeled by (1) and (2); that is, by additive noise and a single drift component. In this case, the expected value in (6) may be evaluated to obtain [1]

$$a(\tau) = \frac{R}{\tau} + \frac{Q_0 \tau}{3}. \tag{8}$$

The first term in this equation is the Allan variance of the noise term $n(t)$, while the second term is the Allan variance of the drift component $b_0(t)$. Because the noise processes are assumed to be statistically independent, the Allan variance of a composite process is equal to the sum of Allan variances of the individual components. On a log–log plot, $a(\tau)$ given by (8) has the shape shown in Fig. 1. An Allan variance plot of a signal consisting only of additive noise and the random drift term $b_0(t)$ has a well-defined minimum. The numerical values of R and Q_0 determine the location of the minimum but the shape of the plot near the minimum

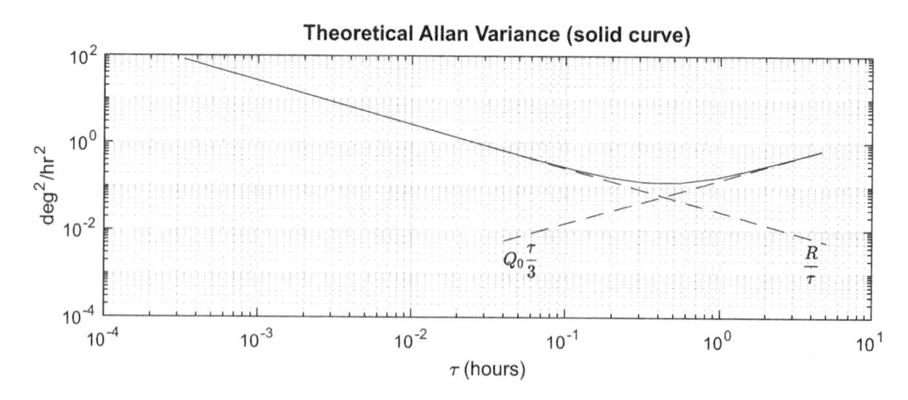

Fig. 1 The theoretical Allan variance (solid curve) for a sensor whose calibration signal consists of additive noise and a single drift component. The Allan variances of the individual components are shown with dashed lines. The spectral densities are $R = 2.65 \times 10^{-2}$ and $Q_0 = 4.31 \times 10^{-1}$ using the gyro units from Table 1

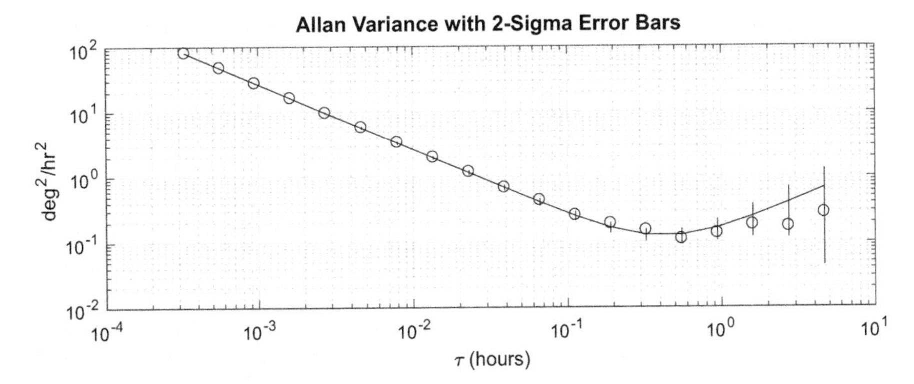

Fig. 2 The theoretical Allan variance of a sensor (solid curve) as well as Allan variance estimates (circles) from 48 h. The vertical bars are centered on the theoretical curve, and range plus or minus two standard deviations, given by the square root of the variance formula, Eq. (9). The vertical bars do not appear to be centered due to the logarithmic scale on the vertical axis. The spectral densities in this simulation are $Q_0 = 4.31 \times 10^{-1}$ and $R = 2.65 \times 10^{-2}$

is independent of the values of R and Q_0. In particular, the plot reaches twice its minimum value at the endpoints of an approximately decade-long interval centered at the minimizing value of τ.

Notice from Fig. 1 that, for values of τ more than one decade below the minimizing value of τ, the Allan variance plot is identical with the line due to the noise term, which depends only on R. For values of τ more than one decade above the minimum, the Allan variance plot is identical with the line due to the drift term, which depends only on Q_0.

The variance of $a(\tau)$, denoted $v(\tau)$, has been computed in [2] to be

$$v(\tau) = \frac{3D - 4\tau}{\tau(D-\tau)^2} R^2 + \frac{(9D - 10\tau)\tau^3}{36(D-\tau)^2} Q_0^2. \tag{9}$$

The square root of (9) gives the standard deviation of the Allan variance for any value of τ. The standard deviation is a measure of the variation in Allan variance points computed from a signal of finite duration D. A typical result is shown in Fig. 2, where the theoretical Allan variance is shown, along with estimates computed from a signal of duration $D = 48$ h. The vertical lines through each computed value are plus or minus two standard deviations. It can be seen that the standard deviations of the computed Allan variance points are small for values of τ to the left of the minimum, and increase rapidly for values of τ to the right of the minimum.

This observation motivates a simple way to estimate the values of R and Q_0 from an Allan variance plot; namely, estimate R by fitting a line with slope -1 on a log–log plot for small values of τ, and estimate Q_0 by fitting a line with slope 1 for large values of τ. This procedure gives reasonable estimates for R but may give erroneous estimates for Q_0. The reason is that the Allan variance points for large values of τ, calculated from a finite amount of calibration data, may be far from the theoretical line (see Fig. 2). Even in this simple case with a single drift component, it is necessary to use a statistical approach to accurately estimate Q_0 [2, 6].

2.3 Multiple Drift Components

Consider the Allan variance plot in Fig. 3. This plot does not have the same well-defined minimum as that in Fig. 1, whose width is a decade in τ (see discussion at the beginning of Sect. 2.2). For the curve in Fig. 3, the width of the interval over which the Allan variance reaches twice its minimum value is more than two decades in τ. The presence of this broad minimum is typical of modern MEMS sensors and indicates the need for additional drift components to model the calibration signal.

The additional components that are needed to model the "flat spot" on the Allan variance plot may all be chosen to be Gauss–Markov (GM) processes with appropriate parameters. The Allan variance plot for a single GM component is shown in Fig. 4. The peak of this plot may be used to "push up" the minimum of

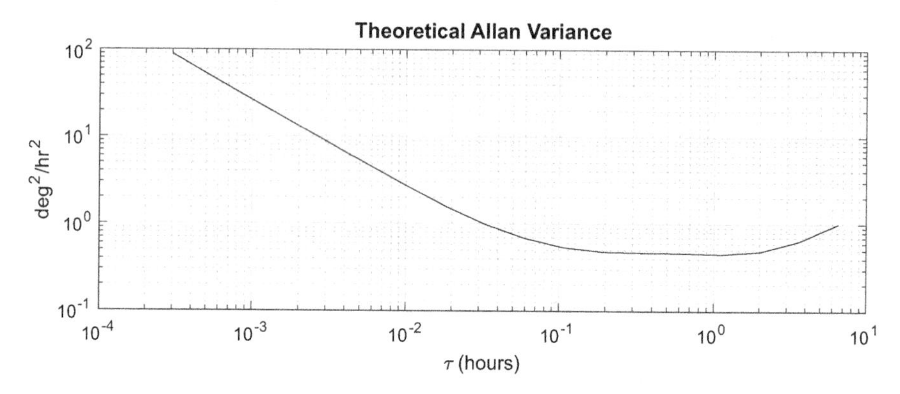

Fig. 3 A theoretical Allan variance plot of a sensor whose statistical model requires more than a single drift component

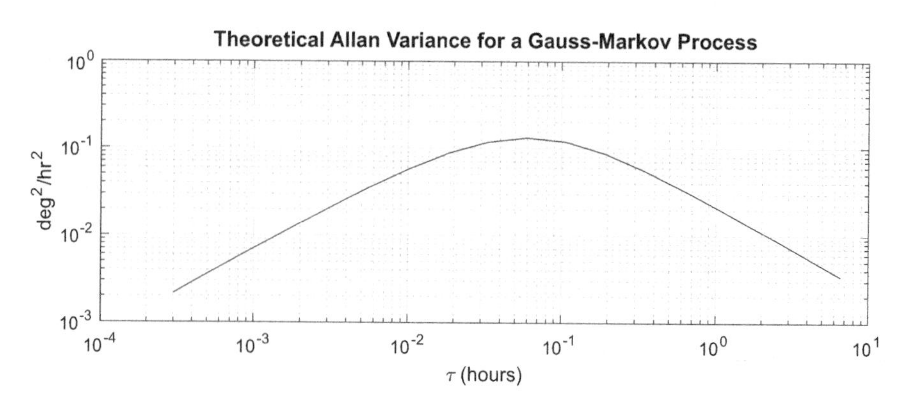

Fig. 4 The theoretical Allan variance of a single GM component described by the parameters $\alpha_1 = 31.6$ and $Q_1 = 21.6$ (see (3))

Table 2 Parameters for the sensor calibration signal whose Allan variance is shown in Fig. 3. See (1)–(3)

Signal component	Parameter(s)
$n(t)$	$R = 2.7 \times 10^{-3}$ deg^2/h
$b_0(t)$	$Q_0 = 4.3 \times 10^{-1}$ deg^2/h^3
$b_1(t)$	$a_1 = 31,\ Q_1 = 22$ deg^2/h^3
$b_2(t)$	$a_2 = 3.6, Q_2 = 5.8$ deg^2/h^3

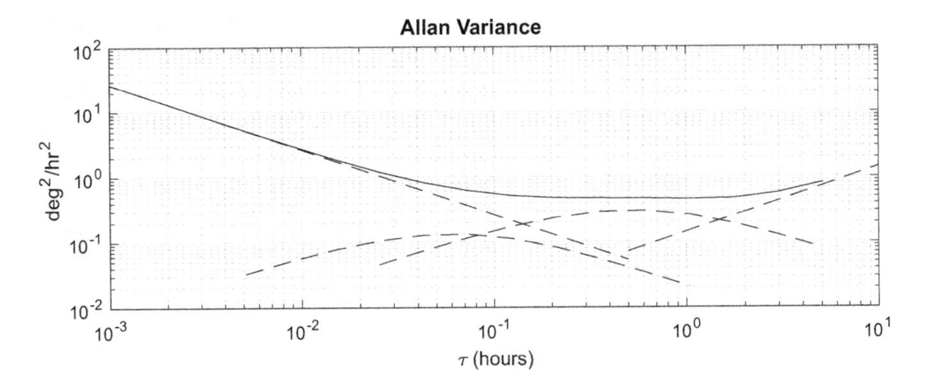

Fig. 5 The theoretical Allan variance plot (solid line) for a sensor whose calibration signal is described by the parameters in Table 2. The dashed lines show the Allan variances of the additive noise and the individual drift components $b_0(t)$, $b_1(t)$, and $b_2(t)$

an Allan variance plot such as that shown in Fig. 1 to create a broad, flat minimum. The Allan variance of a single GM component with parameters (α_i, Q_i) (see (2)) is given by [1]:

$$a(\tau) = \frac{Q_i}{a_i^2 \tau}\left[1 - \frac{1}{2a_i\tau}(3 - 4e^{-a_i\tau} + e^{-2a_i\tau})\right]. \tag{10}$$

By analyzing (10), the following results are obtained for the Allan variance of a single GM component.

$$\text{Peak center occurs at } \tau \approx \frac{1.9}{a_i}, \quad \text{Peak height } \approx 0.19\frac{Q_i}{a_i}. \tag{11}$$

The parameters for the sensor calibration signal whose Allan variance is shown in Fig. 3 are given in Table 2. Figure 5 shows the Allan variance plot for this sensor along with the theoretical Allan variance plots of the additive noise and individual drift components.

It is useful to develop guidelines for choosing the number and approximate parameters of the GM components needed to match a computed Allan variance plot for a given sensor. These guidelines are useful because the parameters of a sensor model are estimated by minimizing a cost function that measures how well a hypothesized model fits a set of Allan variance points computed from a calibration signal (see Sect. 4). The minimization starts with initial guesses of the parameter values, which are obtained using the following guidelines.

To describe the guideline for choosing the number of GM components, let W be the width in decades of the interval $[\tau_1, \tau_2]$ over which the given Allan variance plot reaches twice its minimum value. The number of GM components is then chosen to be the smallest integer greater than W-1. This guideline reflects the fact that the width of the Allan variance peak for a GM component is about a decade, as is the width of the Allan variance minimum for the signal $n(t) + b_0(t)$.

The consideration for choosing the initial guesses of the GM parameters is to uniformly space the GM peaks along the W-interval defined above, with the peak heights set equal to the Allan variance given by the minimum of the sensor's Allan variance plot. To space the peak locations of the C GM components uniformly on a logarithmic scale from τ_1 to τ_2, we first calculate the spacing

$$x = \frac{1}{C+1}(\log_{10}(\tau_2) - \log_{10}(\tau_1)). \tag{12}$$

The τ values of the peak centers, $p_1, \cdots, p_C$ are then given by:

$$p_k = \tau_1 10^{kx}, \; k = 1, \cdots, C. \tag{13}$$

As an example, consider the Allan variance plot shown in Fig. 3. It is seen that W is the interval from $\tau_1 = 4 \times 10^{-2}$ to $\tau_2 = 5.5$, which is a little more than two decades. Thus, we choose the number of GM components to be $C = 2$. Using (12) and (13), we compute the peak centers to be $p_1 = 0.207$ and $p_2 = 1.07$. Figure 6 shows the Allan variance plot for this sensor with τ_1, τ_2, p_1, and p_2 identified on the plot. The minimum of the Allan variance plot in Fig. 3 is about 0.45. Thus, using (11) we have the following initial values for the GM parameters:

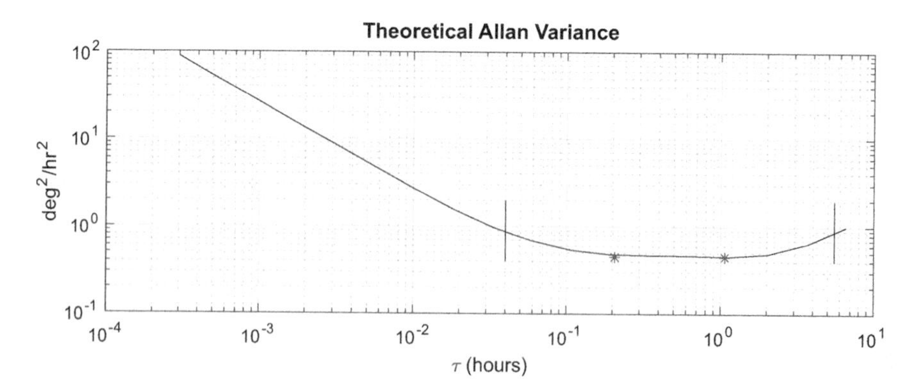

Fig. 6 Allan variance plot for the signal parameters given in Table 2. The vertical lines indicate τ_1 and τ_2, the endpoints of the interval W over which the Allan variance reaches twice its maximum value. The stars indicate the values of p_1 and p_2, which are initial values for the peak centers of two Gauss–Markov processes. See Eqs. (11)–(13)

$$\begin{aligned} a_1 &= \frac{1.9}{p_1} = 9.2 \\ a_2 &= \frac{1.9}{p_2} = 1.8 \\ Q_1 &= \frac{0.45a_1}{0.19} = 21.8 \\ Q_2 &= \frac{0.45a_2}{0.19} = 4.26 \end{aligned} \quad (14)$$

3 Integral Error Bounds

One of the principal uses of rate gyros is to provide an estimate of angular position obtained by integrating the gyro output signal. The presence of additive noise and bias terms in the gyro signal causes errors in the estimated position. It is useful to have a bound on the position error as a function of time. Similarly, with accelerometers, it is useful to have bounds on the first and second integrals of an accelerometer signal.

Bounds on the integral of a sensor output may be obtained by direct calculation as follows. For each component of a sensor calibration signal, obtain an expression for the integral of this component, and compute the variance of the integrated signal as a function of time. A statistical bound on the magnitude of the integrated signal is then given by some multiple of the square root of the variance (standard deviation).

Statistical bounds on the first and second integrals of the $n(t)$ and $b_0(t)$ components were calculated using the direct method in [2, 7]. However, direct calculation for a Gauss–Markov component is tedious. Instead, we derive in the following subsection a relationship between Allan variance and a statistical bound on the integral of a sensor calibration signal.

3.1 *Bound on Single Integral*

The definition of Allan variance given in (6) is independent of k. Thus, we can choose $k = 1$ to obtain

$$a(\tau) = \frac{1}{2}E[(\bar{y}_2 - \bar{y}_1)^2]. \quad (15)$$

Note that $\bar{y}_1$ and $\bar{y}_2$ (see (5)) may be written as

$$\bar{y}_1 = \frac{1}{\tau}X, \quad \bar{y}_2 = \frac{1}{\tau}Y. \quad (16)$$

where

$$X = \int_0^{\tau} y(\lambda)d\lambda, \quad Y = \int_{\tau}^{2\tau} y(\lambda)d\lambda \tag{17}$$

are zero-mean random variables. Then, (15) becomes

$$a(\tau) = \frac{1}{2\tau^2}E[(Y - X)^2] = \frac{1}{2\tau^2}\text{var}(Y - X). \tag{18}$$

Under the assumption that each component of a sensor calibration signal is derived from an independent white-noise process, the random variables are uncorrelated and the variance of $Y - X$ is equal to $\text{var}(X) + \text{var}(Y)$. Furthermore, because X and Y are both obtained from length-τ intervals of data, $\text{var}(X) = \text{var}(Y)$. Thus, (18) may be rewritten as

$$a(\tau) = \frac{1}{2\tau^2} \cdot 2\text{var(X)} = \frac{1}{\tau^2}\text{var}(\theta(\tau)) \tag{19}$$

where $\theta(\tau)$ is the integral of $y(t)$ from 0 to τ. From (19), we have

$$\text{var}(\theta(t)) = t^2 a(t), \tag{20}$$

where $a(t)$ is $a(\tau)$, the Allan variance of $y(t)$, with τ replaced by t.

A useful bound on the magnitude of an integrated calibration signal is given by some multiple, m, times the standard deviation of $\theta(t)$. Such a bound will be called the "$m\sigma$" bound. Using (20), this bound on the first integral of a calibration signal is

$$B_1(t) = m\sqrt{t^2 a(t)}, \tag{21}$$

where m is chosen to achieve a desired (small) value for the probability that the integrated sensor output will exceed this bound. For example, if the noise terms in the calibration signal model have a Gaussian distribution at each point in time, then (21) with $m = 4$ gives the 4σ bound. The probability that the integrated signal exceeds this bound is less than 6.3×10^{-5}, which is the area under a Gaussian probability density function for points exceeding four standard deviations.

For a calibration signal consisting of additive noise and three bias components, the Allan variance $a(\tau)$ is obtained by summing (8) and (10). Changing τ to t and using (21) gives the bound

$$B_1(t) = 4\sqrt{Rt + Q_0 t^3/3 + \text{G}(t)} \tag{22}$$

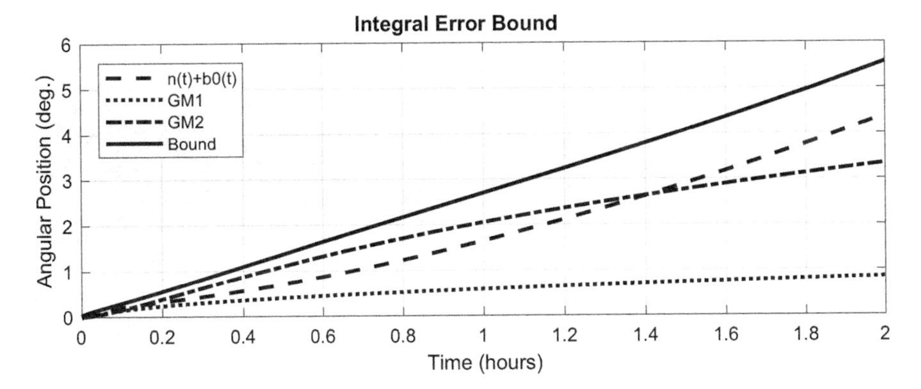

Fig. 7 Error bound $B_1(t)$ (solid curve) from (22) with $m = 4$. Dashed curves are for the individual GM process as well as the combined ARW/RRW component

where $G(t)$ is the contribution of all of the Gauss–Markov terms.

$$\mathrm{G}(t) = \sum_{i=1}^{C} \frac{Q_i}{a_i} \left[\frac{t}{a_i} - \frac{1}{2}(3 - 4e^{-a_i t} + e^{-2a_i t}) \right]. \tag{23}$$

Figure 7 shows the integral bound (22) with $m = 4$ for the calibration signal corresponding to the parameters in Table 2. In addition to the overall bound, individual bounds are also shown for the following components: $n(t) + b_0(t)$, $\mathrm{G}_1(t)$, and $\mathrm{G}_2(t)$. For the first hour, all three of these components contribute significantly to the overall bound. For times greater than 2 h, the contribution of $n(t) + b_0(t)$ begins to dominate, due to $b_0(t)$.

It is important to note that the integral bound given in (21) is a function only of the Allan variance. Thus, if the Allan variance is estimated for some value of the smoothing interval τ, that Allan variance point may be used to obtain a bound on the integrated sensor output at time $t = \tau$. In other words, the integral bound may be obtained directly from computed Allan variance points without the need to estimate the statistical parameters of the sensor. However, the benefit of estimating the parameters is that they may be used to obtain a bound on the double integral of the sensor output, as is shown next.

3.2 Bound on Double Integral

If a bound on the double integral of a sensor calibration signal is required, then it is necessary to estimate the statistical parameters $[R, Q_0, (\alpha_1, Q_1), \cdots, (\alpha_C, Q_C)]$ shown in (22) and (23). The bound $B_2(t)$ on the second integral of a calibration signal is simply the integral of the first-integral bound:

$$B_2(t) = \int_0^t B_1(\lambda) d\lambda. \tag{24}$$

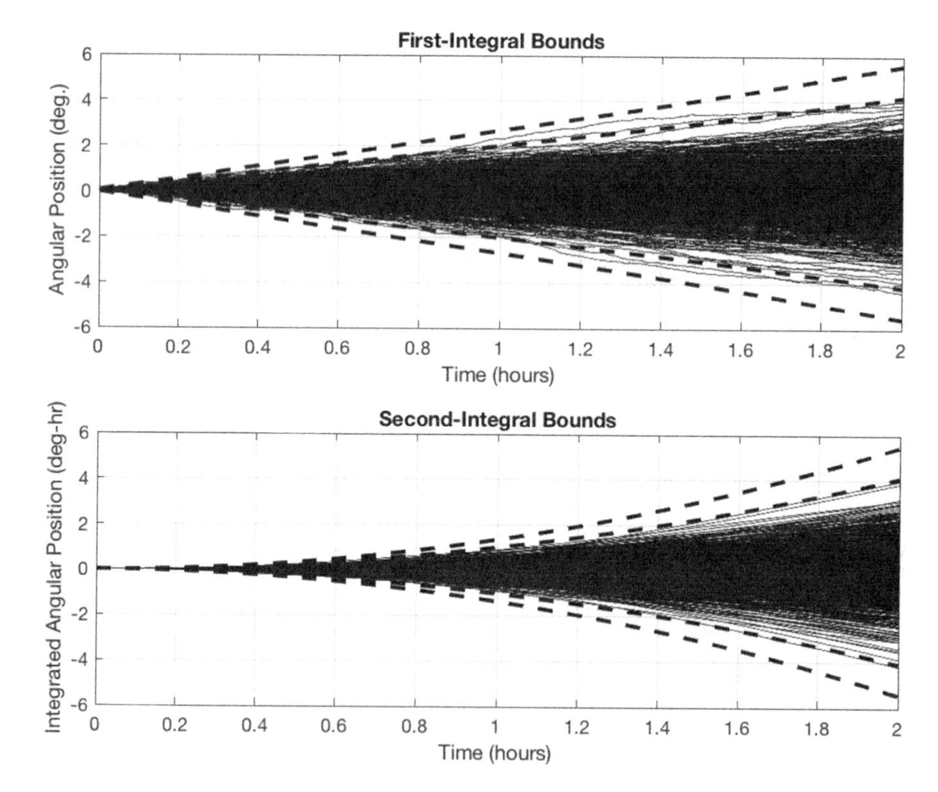

Fig. 8 Bounds on the first integral (top graph) and second integral (bottom graph) for the sensor whose parameters are given in Table 2. In both graphs, the outer dashed lines are the 4σ $(m = 4)$ bound and the inner dashed lines are the 3σ bounds. The solid lines are the first and second integrals of 1000 simulated calibration signals

After the model parameters are estimated, the first-integral bound can be evaluated on a dense grid of time points and the integral in (24) may be evaluated numerically using (22) and (23). Figure 8 shows the first- and second-integral bounds for the sensor whose parameters are given in Table 2, as well as the first and second integrals of 1000 simulated calibration signals.

4 Estimating the Parameters of a Sensor Model

The integral bounds given in the previous section are a function of the parameters describing the noise and drift components of the sensor. These parameters may be estimated by first computing the Allan variance from a finite record of calibration data, and then fitting a model to the experimental Allan variance points. One way to do this is by a simple least-squares fit. Let $\boldsymbol{\tau}$ be a vector of τ values. Let $\hat{a}(\boldsymbol{\tau})$ be the corresponding vector of computed Allan variance points, and let $a(\boldsymbol{\tau}; \boldsymbol{\theta})$ be

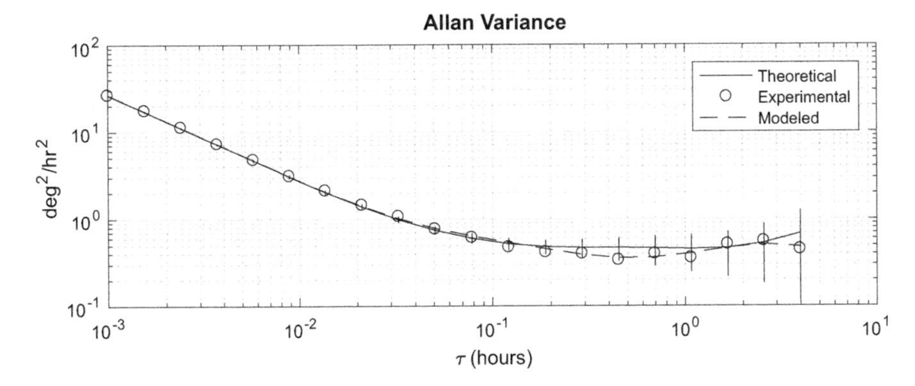

Fig. 9 Allan variance for the sensor whose parameters are given in Table 2. The solid line is the theoretical Allan variance. Experimental Allan variance points were computed from a calibration signal of duration $D = 48$ h. Model parameters obtained by a least-squares fit to the experimental Allan variance points were used to construct the dashed line. The vertical lines indicate plus or minus two standard deviations for each computed Allan variance point

the vector of theoretical Allan variance points given by the sum of (8) and (10) for some values of the model parameters $\boldsymbol{\theta} = [R, Q_0, (a_1, Q_1), \cdots, (a_C, Q_C)]$. The least-squares estimate of the parameter vector is

$$\hat{\boldsymbol{\theta}}_{LS} = \arg\min_{\boldsymbol{\theta}} (\hat{a}(\boldsymbol{\tau}) - a(\boldsymbol{\tau}; \boldsymbol{\theta}))^T (\hat{a}(\boldsymbol{\tau}) - a(\boldsymbol{\tau}; \boldsymbol{\theta})). \tag{25}$$

A method to obtain initial values of the parameters in $\boldsymbol{\theta}$ is described in Sect. 2. When the minimization has been performed, the estimated parameters in $\hat{\boldsymbol{\theta}}_{LS}$ may be used in (8) and (10) to calculate a modeled Allan variance for any value of τ.

Note that, from a statistical point of view, $\hat{\boldsymbol{\theta}}_{LS}$ is not the best estimator of $\boldsymbol{\theta}$. Indeed, [6] has shown that $\hat{\boldsymbol{\theta}}_{LS}$ is not, in general, a consistent estimator of $\boldsymbol{\theta}$. Nevertheless, for values of τ at which the calculated Allan variance has small standard deviation, the least-squares fit will be close to the theoretical Allan variance. See, for example, Fig. 9 for values of τ less than 2 h.

A better estimate of theta, $\hat{\boldsymbol{\theta}}_{WLS}$ is obtained by a weighted least-squares fit, and was derived in [8, 9] and given the name Generalized Method of Wavelet Moments (GMWM). This estimator is

$$\hat{\boldsymbol{\theta}}_{WLS} = \arg\min_{\boldsymbol{\theta}} (\hat{a}(\boldsymbol{\tau}) - a(\boldsymbol{\tau}; \boldsymbol{\theta}))^T \mathbf{W}^{-1} (\hat{a}(\boldsymbol{\tau}) - a(\boldsymbol{\tau}; \boldsymbol{\theta})) \tag{26}$$

where $\mathbf{W}$ is a diagonal matrix whose diagonal elements are the variances of the computed Allan variance points. A formula for estimating these variances from a single calibration signal is given in [8] and [10].

The performance of the weighted least-squares estimator, for a single calibration signal, is shown in Fig. 10, which is identical to Fig. 9 except that the modeled Allan variance is computed using $\hat{\boldsymbol{\theta}}_{WLS}$. Comparing Fig. 10 with Fig. 9, it can be seen that

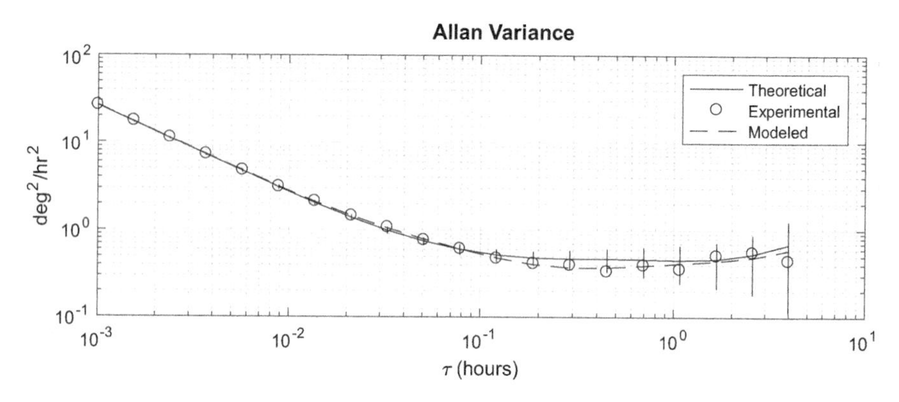

Fig. 10 Allan variance for the sensor whose parameters are given in Table 2. The solid line is the theoretical Allan variance. Experimental Allan variance points were computed from a calibration signal of duration $D = 48$ h. Model parameters obtained by a least-squares fit to the experimental Allan variance points were used to construct the dashed line. The vertical lines indicate plus or minus two standard deviations for each computed Allan variance point

both modeled curves are close to the true Allan variance for values of τ less than 2 h. The two modeled curves differ from each other only for the largest values of τ, where the dashed line from the least-squares parameters (Fig. 9) curves downward, while the dashed line for the weighted least-squares parameters (Fig. 10) curves upward.

Table 3 shows the parameter values given by the least-squares and weighted least-squares estimators for a single calibration signal. It can be seen that the two sets of parameters are quite different from each other, and that the least-squares estimate of the parameter Q_0 is a negative number, which is not physically meaningful. This reflects the fact that the weighted least-squares estimator is the superior estimator.

From the point of view of integral error bounds, however, it is the numerical value of the Allan variance that matters, not the particular parameter values used in the model. If the integral error bounds are needed only over a small interval of time, say less than 2 h, then these bounds are a function of the modeled Allan variance for $\tau < 2$. For the example given in this section, the least-squares and weighted least-squares give nearly identical modeled Allan variance for $\tau < 2$ h. Thus, both methods give nearly identical integral error bounds up to 2 h.

Figure 11 shows the integral error bounds obtained from least-squares parameter estimates from 1000 calibration signals along with (dark lines) the error bounds computed from the true parameter values. Plots of the integral error bounds obtained from the weighted least-squares parameter estimates look identical to Fig. 11. This figure indicates that the error bounds computed from parameter estimates obtained from a calibration signal are random variables at each moment in time, whose means are the theoretical error bounds, and whose variances grow with time.

Table 3 Parameter values given by the least-squares and weighted least-squares estimators ((25) and (26), respectively) computed from a calibration signal of 48-h duration

Parameter	True value	LS	WLS
R	2.7×10^{-3}	2.7×10^{-3}	2.7×10^{-3}
Q_0	4.3×10^{-1}	-3.5×10^{-1}	3.3×10^{-1}
a_1	3.1×10^{1}	2.7×10^{1}	4.6×10^{1}
Q_1	2.2×10^{1}	1.9×10^{1}	4.0×10^{1}
a_2	3.6×10^{0}	1.75×10^{0}	2.5×10^{0}
Q_2	5.8×10^{0}	3.0×10^{-1}	1.9×10^{0}

The models were fit to the same experimental Allan variance points shown in Figs. 9 and 10

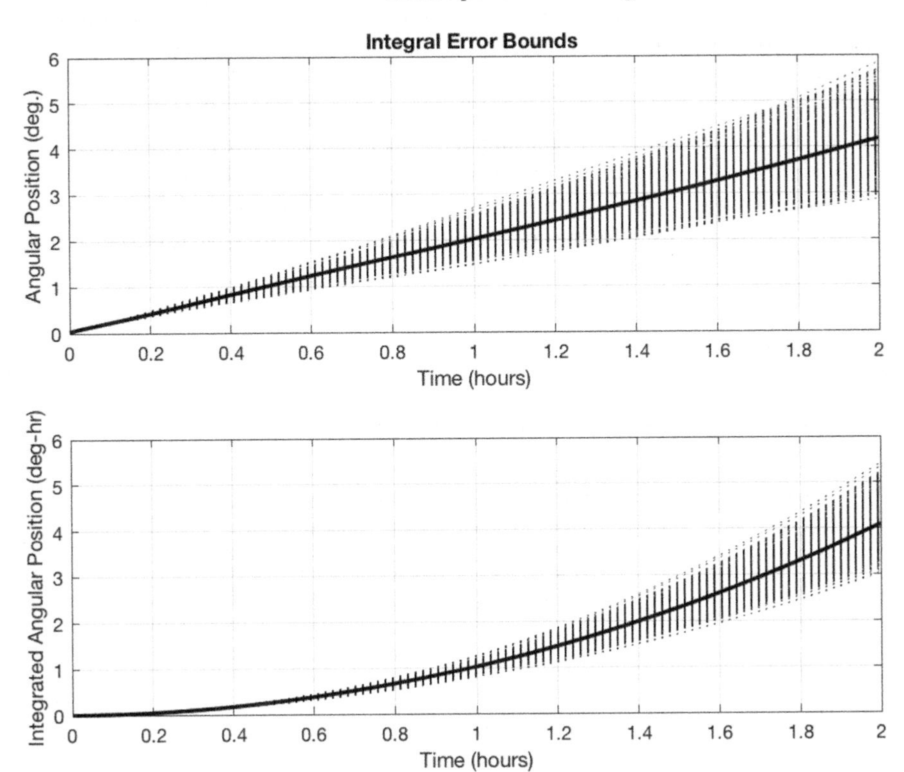

Fig. 11 3σ bounds on the first integral (top graph) and second integral (bottom graph) for the sensor whose parameter values are given in Table 2. In both graphs, the dark curve is the bound computed using the true parameter values and the dotted curves are the bounds computed from least-squares parameter estimates obtained from 1000 different calibration signals. The 1000 dotted curves appear as vertical lines

5 Summary and Future Work

The main contribution of this chapter is the derivation of a statistical bound on the magnitude of an integrated calibration signal. This bound, given by (21), is a simple function of the Allan variance of a sensor. Thus, the bound may be computed directly from experimental Allan variance points without the need to estimate model parameters. The resulting bound is useful when the variance of the calculated Allan variance points is small.

In order to compute the bound on the second integral, the first-integral bound must be evaluated on a dense grid of points for accurate numerical integration. A straightforward way to do this is to estimate parameters for a model of the sensor calibration signal and evaluate the theoretical Allan variance of the model on a dense grid of points. This was demonstrated in Sect. 4.

An area of future work has to do with obtaining integral error bounds from a single calibration signal even when the variance of the calculated Allan variance points is not small. It was seen in Fig. 11 that the first-integral bound computed from a calibration signal is a random variable whose mean is the theoretical bound and whose variance is related to the variance of the estimated Allan variance. This can also be seen from (21) by replacing $a(t)$ with an estimate $\hat{a}(t)$ obtained from a calibration signal. The variance of $\hat{a}(t)$ could be used to find the variance of $B_1(t)$, which could then be used to find a first-integral bound from a single calibration signal that would be valid for the complete range of calculated Allan variance points.

Appendix

Program to Simulate Calibration Signal

```
% Script to compute calibration signal and compute
% and plot Allan variance
T=1/100/3600; % sampling interval (hours)
D=48; % duration of calibration signal (hours)
N=floor(D/T);
% Define parameters for sensor
Q=4.3047e-01;
R=2.6518e-02;
a1=3.0957e+01;
Q1=2.1572e+01;
a2=3.6367e+00;
Q2=5.7928e+00;
arw=randn(N,1)*sqrt(R/T);
rrw=randn(N,1)*sqrt(Q1/T);
rrw2=randn(N,1)*sqrt(Q2/T);
rrw3=randn(N,1)*sqrt(Q*T);
A=[1 -exp(-a1*T)];
B=(1-exp(-a1*T))/a1;
frrw=filter(B,A,rrw);
A2=[1 -exp(-a2*T)];
```

```
B2=(1-exp(-a2*T))/a2;
frrw2=filter(B2,A2,rrw2);
y=arw+frrw+frrw2+cumsum(rrw3);
% Compute Allan variance
tau=logspace(-3,0.6,20)'; %smoothing lags in hours
m=floor(tau/(T)); %smoothing lags in samples
[avar]=compute_avar(y,m);
loglog(tau,avar)
```

Program to Compute Allan Variance

```
function avar=compute_avar(y,m)
% y is a calibration signal
% m is a vector of smoothing interval lengths
M=floor(length(y)./m);
jj=length(m);
avar=zeros(jj,1);
for j=1:jj
    mm=m(j);
    MM=M(j);
    D=zeros(MM,1);
    for i=1:MM;
        D(i)=sum(y((i-1)*mm+1:i*mm))/mm;
    end
    v=diff(D).^2;
    avar(j)=0.5*(mean(v));
end
end
```

Program to Estimate Sensor Parameters by Least Squares

```
function [R,Q,Q1,a1,Q2,a2]=LSfit(avar,tau,x0)
% this function fits a model of additive noise,
random walk,
% and two GM components.  x0 is a vector of initial
guesses
%  for each parameter that the function returns
% avar is a vector of Allan variance points computed
% at each value of tau, which is a vector of
smoothing intervals
options=optimset('display','off','TolX',1e-6,
   'TolFun',1e-6);
```

```
x=fminsearch(@efun,x0,options,tau,avar);
R=x(1);
Q=x(2);
Q1=x(3);
a1=x(4);
Q2=x(5);
a2=x(6);
function f=efun(x,tau,avar)
R=x(1);
Q=x(2);
Q1=x(3);
a1=x(4);
Q2=x(5);
a2=x(6);
x1=R./tau;
x2=Q*tau/3;
x3=Q1/a1^2./tau.*(1-1/2/a1./tau.*(3-4*exp(-a1*tau)
+exp(-2*a1*tau)));
x4=Q2/a2^2./tau.*(1-1/2/a2./tau.*(3-4*exp(-a2*tau)
+exp(-2*a2*tau)));
%err=(log10(avar)-log10(x1+x2+x3+x4));
err=((avar)-(x1+x2+x3+x4));
f=sum(err.*err);
```

References

1. *IEEE Standard Specification Format Guide and Test Procedure for Single-Axis Interferometric Fiber Optic Gyros,* IEEE Std. 952–1997, 1998
2. R.J. Vaccaro, A.S. Zaki, Statistical modeling of rate gyros. IEEE Trans. Instrum. Meas. **61**(3), 673–684 (2012)
3. J.J. Ford, M.E. Evans, Online estimation of Allan variance parameters. J. Guid. Control Dyn. **23**(1), 980–987 (2000)
4. R.J. Vaccaro, A.S. Zaki, Reduced-drift virtual gyro from an array of low-cost gyros. Sensors **17**(2), E352 (2017). https://doi.org/10.3390/s17020352
5. N. El-Sheimy, H. Hou, X. Niu, Analysis and modeling of inertial sensors using Allan variance. IEEE Trans. Instrum. Meas. **57**(1), 140–149 (2008)
6. S. Guerrier, R. Molinari, Y. Stebler, Theoretical limitations of Allan variance-based regression for time series model estimation. IEEE Signal Process Lett. **23**(5), 595–599 (2016)
7. R.J. Vaccaro, A.S. Zaki, Statistical modeling of rate gyros and accelerometers, in *Proceedings of IEEE/ION PLANS 2012*, Myrtle Beach, SC, 23–26 April 2012
8. S. Guerrier, J. Skaloud, Y. Stebler, M. Victoria-Feser, Wavelet-variance-based estimation for composite stochastic processes. J. Am. Stat. Assoc. **108**(503), 1021–1030 (2013)
9. S. Guerrier, R. Molinari, Y. Stebler, Wavelet-based improvements for inertial sensor error modeling. IEEE Trans. Instrum. Meas. **65**(12), 2693–2700 (2016)
10. D.B. Percival, A.T. Walden, *Wavelet Methods for Time Series Analysis* (Cambridge University Press, New York, 2000)

Developing Efficient Random Flight Searches in Bounded Domains

Thomas A. Wettergren

1 Introduction

The search for hidden objects is one of the most basic problems of behavioral ecology, as it is a fundamental process in foraging for food sources. In that light, there is certain biological interest in understanding the decision process that animals use to achieve foraging success with the least effort possible [15, 16]. This is similar to the desired decision paradigm for any entity performing search, which is why the replication of foraging search in engineered systems has been an active area of study [2, 3]. Foraging search is particularly attractive as a mechanism when the search agent needs to behave independently, with no external (or supervisory) control. In those cases, developing search strategies for an engineered system that mimic the search process of a biological forager can be beneficial. Furthermore, the mathematical investigation of foraging search provides insight into the process that is being optimized by the forager, and the resulting optimal decision rule can provide insight into engineering man-made systems for search.

When classic search theory was developed in the 1940s, it was employed for manned assets that controlled their motion and attempted to maximize encounters with objects of interest [8]. This type of search planning led to Koopman's *random search formula*, which shows that the probability of finding an object over search time t is given by $p(t) = 1 - \exp(-SE \cdot t)$. In this formulation, SE is the incremental search effort applied over a unit time step, which is nominally given by the fraction of the search space covered in a unit time step multiplied by the searcher's probability of detection. Koopman's random search formula is used in

T. A. Wettergren (✉)
Naval Undersea Warfare Center, Newport, RI, USA
e-mail: thomas.wettergren@navy.mil

A. A. Ruffa, B. Toni (eds.), *Advanced Research in Naval Engineering*, STEAM-H: Science, Technology, Engineering, Agriculture, Mathematics & Health, https://doi.org/10.1007/978-3-319-95117-1_10

most search planning activities, as it accounts for the probabilistic nature of the search process. However, the use of this search formula implies parallel search tracks that attempt to "cover" the search region, in what is formally known as *boustrophedon search* or more colloquially known as *mow-the-lawn search*. Animal foraging search, in contrast to this highly structured coverage search, does not try to completely cover an area in finding prey, but instead seeks to find the prey as quickly and efficiently as possible. For this reason, foraging search tends to focus on a different modeling paradigm than the coverage search found in most of the engineering search literature.

When dealing with multiple foragers trying to find prey in the same area, the problem becomes one of dividing search effort given a scarce resource. This is the mathematical structure under which most foraging studies are taken. There has been a popular view in the mathematical ecology community that simple two-player games are reasonable representations of behavior found in animal conflict [15]. When foraging for prey as a resource, this search for scarce resources has been viewed as a game between foragers and the randomly hidden resources (prey) [10]. Andersson [1] has shown that animal populations can be mathematically analyzed using computational optimization of the search parameters. In that context, the search activity of foragers has been modeled as an optimization of the specific parameters in a search game. Using this modeling paradigm, it is clear that an optimal search can be conducted that is tailored to the amount of a priori information available about both the disposition of the object of search as well as the expected search capability of the searcher in various subdomains of the search region.

Using this idea of foraging as a search process that plays out as a game between searchers and the prey, we formulate the search for hidden objects by a searcher who performs with limited guidance as a foraging model. Another feature of ecological search that we exploit is the use of random search as a search paradigm when there is uncertainty in the environment and the performance of the searcher. Whereas random walks may seem to be a natural choice for random exploration of a space, it has been observed that Lévy flights are actually encountered regularly in nature as a more common random search strategy [12, 18].

While most of the ecological search problems are effectively performed on an unbounded domain (i.e., animals with free range to find prey), engineered systems are typically employed to perform search in a specifically restricted domain. The limited work that has been performed on flight searches in bounded domains [22] has focused on observing the effect of the bounded domain on flights designed without regard for domain size. In this work, we use the biological paradigm as inspiration to optimally design a Lévy flight-based search for a searcher performing search in a bounded domain. We specifically design a method for selecting the parameters of an optimized foraging-style random search for objects and show its effectiveness in bounded domains. We also include a strategy for modification of these parameters for problems with multiple searchers in the same domain. Numerical examples are included to show the performance of these searches compared to nonrandom search strategies. We conclude with some guidance on the types of engineering problems where this new random search theory could be beneficially applied.

2 Flight Models of Biological Foraging

Many species of animals perform a foraging process that unfolds as a random search for prey [16]. While many engineered systems may look at search as a process of following a prescribed pattern of motion in a domain, animals do not have that luxury. This happens for multiple reasons. First, the animals do not have prior guidance on the details of the search domain, they typically only have a feel for the "patchiness" of the search domain, or the expected density of prey sites that can be found. Because of this, there are no a priori planned paths for conducting a search. Secondly, animals tend to have limited inertial navigation capability, which causes a difficulty in following a prescribed path. Thirdly, the search performance of an animal in a domain (i.e., their effective range for spotting prey) is often variable and only known in real time. This causes a need for adaptivity throughout the plan. All of these considerations lead to a natural consequence that animals perform randomized search strategies while foraging. Since many of these considerations also exist for engineered systems operating in difficult environmental conditions, it is of interest to understand how to optimize the performance of these types of foraging strategies for potential use in engineered systems that are designed to perform search.

The specific spatial search patterns that animals follow during their foraging behaviors have been accurately modeled as a random flight process (a random walk with variable step sizes) when the flight leg lengths follow a Lévy distribution [19, 20]. This style of flight process has a number of short legs with intermittent long legs, leading to searches that focus on a small area for a period of time and then occasionally make a long step to get to a new area (randomly selected) for search. The justification that has been developed for modeling with a Lévy flight is that the Lévy flight model is a consequence of modeling the foraging interaction of searchers and resources as a random search [12], and it has been further shown that a deterministic random walk in an uncertain random environment provides Lévy-like encounters [5]. While it may at first appear that a Brownian random walk would provide better results when searching for random objects, it has been shown that the "patchiness" of the expected resources (due to the environmental context) affects the bias of the random walk towards the Lévy pattern becoming optimal [7]. In particular, it has been shown that optimal random motion search strategies tend from Brownian to Lévy when the resource spatial density is small (relative to the foragers field of view) and the searchers move faster than the resources that they seek [4]. However, there is some recent evidence that in certain cases Brownian search can beat Lévy search, even for sparse problems [11]. Furthermore, when target sites are revisited shortly after initial contact, there has been some evidence [14] of a transition from Lévy search to ballistic random walks.

The key observation that kicked off much of the work in Lévy search models of biological foraging was that foragers followed random flights whose leg lengths followed distributions that had "heavy tails" [18]. When these heavy-tailed distributions are applied to select the leg lengths of a random flight model, the model becomes one of the *superdiffusive* random flights. In a typical normal diffusion process (such as a Brownian flight), the mean square displacement of a particle (e.g., the searcher in our search context) grows linearly with time. This leads to a search rate that grows as $\sqrt{t}$ (as it is proportional to the square root of the mean square displacement). However, for a superdiffusive diffusion process (such as a Lévy flight) the mean square displacement of a particle follows a superlinear relationship with time (typically close to quadratic). Thus, for a search based upon a superdiffusive random flight, the search rate grows as t^{α} with $\alpha > 1/2$, and as the mean square displacement approaches a quadratic function of time the search rate grows nearly linearly. This improvement in search rate is hypothesized as the primary reason why biological systems apply superdiffusive random flight as opposed to other diffusive processes. We note that a linear search rate is the best rate that can be achieved by any fixed-speed searcher, thus the superdiffusive flight is the best chance for a random flight to approach optimal search rates.

Since superdiffusive random flight is the overarching goal of a random searcher, we propose to use a random flight model with leg lengths chosen from the following distribution (similar to the model in [22]):

$$p(\ell) = \begin{cases} 0, & \ell < \ell_0 \\ c\,\ell^{-\mu}, & \ell \geq \ell_0 \end{cases} \tag{1}$$

with a value of $c = (\mu - 1)\ell_0^{\mu-1}$ to make it a valid probability density function. This probability density function can be easily shown to be superdiffusive when the exponent μ is in the range $1 < \mu \leq 3$. As $\mu \to 1$, it approaches ballistic behavior, and for $\mu > 3$ it follows regular diffusion. Because of the desire to maintain superdiffusive behavior, we consider values of $\mu \approx 2$. Equation (1) is the leg length basis of our search paradigm. In practice, a searcher would randomly (uniform random) select a search direction along with a leg length randomly chosen from the distribution in Eq. (1). The search is then conducted in the chosen direction for the chosen length. At the end of the leg, a new direction and a new leg length are randomly chosen and the process repeats until the search concludes (either due to finding the object or reaching the prescribed end-of-search time).

For the bounded domains that are of interest in this work, the resulting legs are truncated whenever they reach the boundary of the domain. In that case, the direction of the next leg is chosen to be random but limited to those directions that point to the interior of the domain. In Fig. 1, we show an example of a random superdiffusive Lévy-style flight performed in a bounded domain. The total length of the search path in this example was chosen to match that of a 10-leg mow-the-lawn search plan for the same domain. Note how the random flight leads to some portions of the domain with significant overlap and other portions of the domain without coverage. Because of these issues with random search, it is anticipated that random search would only be a preferable paradigm in scenarios with sparse coverage (i.e., where only a fraction of the total domain is searched).

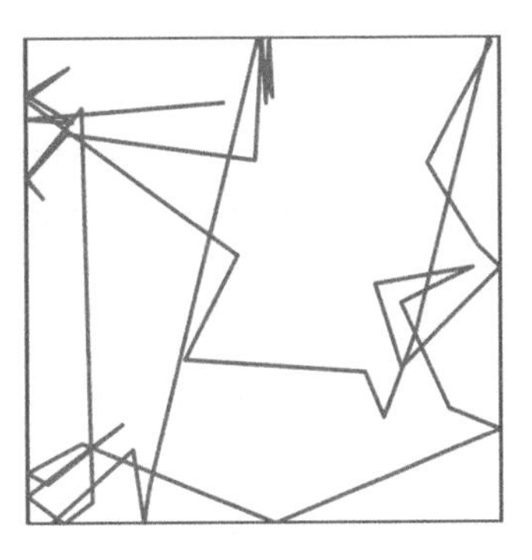

Fig. 1 A long-duration Lévy flight search performed in a bounded domain. The total search distance is equal to that for a uniform mow-the-lawn search plan for a region of identical size

3 Optimizing Search Encounter Rates

The use of a Lévy flight for search, as opposed to a flight following other distributions (such as a simple Brownian motion flight), is based upon a notion of search encounter rate. In particular, when searching for objects, the amount of search effort applied can be described by the amount of the search area that is accumulated in the field of view of the searcher as the searcher moves along. Specifically, this search effort is the integral of the searcher field of view over the motion path taken by the searcher, normalized by the total search area. For any straight search leg from time t_i to t_{i+1}, which is of speed v, this search effort SE_i is given by:

$$SE_i = SE(L_i) = 2r_d L_i / A_0 \tag{2}$$

where r_d is the radius of searcher detection (the field of view of the searcher), $L_i = v(t_{i+1} - t_i)$ is the distance (length) transited by the search leg, and A_0 is the size of the total search area. In the absence of any variation in performance over the search area (such as due to varying environmental conditions), the search effort is a simple linear function of search leg length. Thus, we follow the notation in [12] to define a search efficiency, η, as the ratio of the number of object detections to the total search effort applied by a searcher. Since we can assume that the total distance is the product of the total number of object detections M times the average distance traveled between two successive detections $\langle L \rangle$, we have

$$\eta = \frac{MA_0}{2r_d M \langle L \rangle} = \frac{c}{\langle L \rangle} \tag{3}$$

where $c = A_0/(2r_d)$ is a scaling constant for the geometrical size of the specific problem. From Eq. (3), it is clear that the maximization of search efficiency η is identical to the minimization of $\langle L \rangle$, the distance traveled between two successive detections. Thus, the goal of maximizing search efficiency becomes one of minimizing the average travel distance between successive detections.

Assume that we have reached some point in the search process where a detection of an object has just been made. To then minimize the distance to the next detection, we make the following two observations: (1) a long leg is beneficial only if oriented in the direction of another object, and (2) a short leg is beneficial only for reorienting the searcher away from a "bad" direction. Thus, it is clear that, on average, it is optimal to pick the length of the search leg to match the expected distance to the next object. This expected distance to the next object is given by the mean free path between detection of objects λ_T, which in two dimensions for stationary objects is given by:

$$\lambda_T = \frac{1}{2r_d\rho} \tag{4}$$

where r_d is the distance at which the searcher detects an object, and ρ is the density of objects in the domain. Therefore, we arrive at the conclusion that the optimal search strategy for maximization of search efficiency is to have the average search leg length match the mean free path λ_T. Due to the boundedness of our search domain, many of the longer search legs will become truncated, and thus the average search leg length that is achieved is not merely an average of the leg distribution function given by Eq. (1). In the next section, we derive an approximate analytic expression for this average achieved search leg length and then match it to the mean free path to obtain a rule for setting optimal search parameters for random search flights in bounded domains.

We note that the search encounter rates of this section are only valid for sparse target situations. In particular, it has been ecologically observed [17] that the prey encounter rate and prey density are related by spatial issues leading to more density dependency than a pure frequentist approach (as shown above) suggests. This effect is most pronounced for non-sparse scenarios. Furthermore, it has been also shown that this dependency, when it is a noticeable effect, can be described by an increase in search rate over time [13]. We specifically limit our study to sparse scenarios, and these issues that arise in non-sparse scenarios are a subject of future work.

3.1 Expected Search Leg Lengths

When applying a random flight model to a bounded domain, there is a forced truncation of many of the longer length flight legs, which greatly impacts the resulting distribution of coverage in the domain. To assess the impact, we model the truncation by assuming that, when truncation occurs, the effective leg length is truncated at some length given by the random variable ξ. Let us first analyze the resulting performance for a fixed value of ξ (i.e., when the random variable follows a delta function distribution).

We begin with the following lemma for computing the expected value of a random variable x that is sampled from a known distribution $f(x)$ but then truncated at some maximum value λ:

Lemma 1 *Let $x \in \mathbb{R}^+$ be a random variable with probability density function $f(x) : \mathbb{R}^+ \to [0, 1)$. Let y be another random variable defined through the transformation $y = g(x)$ where $g(x) = \min\{x, \lambda\}$ for a real parameter λ. Then, the expected value of y is given by:*

$$E[y] = \int_0^\lambda x f(x) dx + \lambda \int_\lambda^\infty f(x) dx \tag{5}$$

Proof The expected value of a function $g(x)$ of a continuous random variable $x \in \mathbb{R}^+$ with probability density function $f(x)$ is given by:

$$E[g(x)] = \int_0^\infty g(x) f(x) dx.$$

The range of integration can then be split into two parts for any real parameter $\lambda \in (0, \infty)$ according to:

$$E[g(x)] = \int_0^\lambda g(x) f(x) dx + \int_\lambda^\infty g(x) f(x) dx.$$

Now, from the definition of $g(x)$ it is clear that $g(x) = x$ over the range of the first integral and $g(x) = \lambda$ over the range of the second integral. Therefore, we have

$$E[y] = E[g(x)] = \int_0^\lambda x f(x) dx + \int_\lambda^\infty \lambda f(x) dx,$$

which is the resulting form in the lemma. □

Now, the mean leg length $E[\ell]$ for searches that follow leg length distribution $p(\ell)$, but are truncated at some fixed length $\ell = \xi$, is given (according to Lemma 1) by:

$$E[\ell] = \int_0^\xi \ell \; p(\ell) \, d\ell + \int_\xi^\infty \xi \; p(\ell) \, d\ell \tag{6}$$

By substituting in the distribution for $p(\ell)$ from Eq. (1), this expression simplifies to

$$E[\ell] = (\mu - 1)\ell_0^{\mu-1} \left\{ \int_{\ell_0}^\xi \ell^{1-\mu} \, d\ell + \int_\xi^\infty \xi \; \ell^{-\mu} \, d\ell \right\} \tag{7}$$

Taking the integrals (and recalling that $\mu > 1$) leads to

$$E[\ell] = \left(\frac{\mu - 1}{2 - \mu}\right)\left[\frac{\xi^{2-\mu} - \ell_0^{2-\mu}}{\ell_0^{1-\mu}}\right] + \xi\left(\frac{\xi}{\ell_0}\right)^{1-\mu} \tag{8}$$

which can be algebraically rearranged to yield

$$E[\ell] = \ell_0\left[\left(\frac{1-\mu}{2-\mu}\right) + \left(\frac{1}{2-\mu}\right)\left(\frac{\xi}{\ell_0}\right)^{2-\mu}\right] \tag{9}$$

We note that the expression in Eq. (8) is similar to expressions found by Viswanathan and colleagues [20], but their analysis uses a truncation based upon the termination of legs due to detection events, not due to geometric bounds on the search region. The expression in Eq. (9) is more useful for our purposes than that in Eq. (8), as it demonstrates the deviation from linearity when viewing the mean leg length as a function of the truncation value ξ.

For the practical problem of a bounded domain, we consider the length parameters to be truncated by random lengths due to the relation between the current leg orientation and the domain geometry. For this, we replace the truncation length value ξ with a random variable $\xi_d \in \mathbb{R}^+$ with probability density function $h(\xi_d)$ whose mean value is given by $E[\xi_d] = \int \xi_d h(\xi_d)\, d\xi_d = \lambda_d$. We also define the mean value of the resulting leg lengths to be given by $\langle \ell \rangle$, which is related to $E[\ell]$ by the following:

$$\begin{aligned}\langle \ell \rangle &= \int_0^\infty E[\ell] h(\xi_d)\, d\xi_d \\ &= \left(\frac{1-\mu}{2-\mu}\right)\ell_0 + \left(\frac{1}{2-\mu}\right)\int_0^\infty \left(\frac{\xi_d}{\ell_0}\right)^{1-\mu} \xi_d\, h(\xi_d)\, d\xi_d. \end{aligned} \tag{10}$$

The mean leg length formula of Eq. (10) is valid for values of $1 < \mu \leq 3$. However, for the application of search design, we focus near the center of this range where $\mu \approx 2$. By letting $\varepsilon = 2 - \mu$ for some $-1 \ll \varepsilon \ll 1$, we see that the integral term in Eq. (10) can be approximated by:

$$\begin{aligned}\int_0^\infty \left(\frac{\xi_d}{\ell_0}\right)^{1-\mu} \xi_d\, h(\xi_d)\, d\xi_d &= \ell_0^{\mu-1}\int_0^\infty \xi_d^{\varepsilon}\, h(\xi_d)\, d\xi_d \\ &= \ell_0^{\mu-1}\int_0^\infty (\lambda_d + (\xi_d - \lambda_d))^{\varepsilon}\; h(\xi_d)\, d\xi_d \\ &\approx \ell_0^{\mu-1}\int_0^\infty (\lambda_d^{\varepsilon} + \varepsilon\lambda_d(\xi_d - \lambda_d))\, h(\xi_d)\, d\xi_d \end{aligned}$$

$$\approx \ell_0^{\mu-1} \lambda_d^{\varepsilon} \int_0^{\infty} h(\xi_d)\, d\xi_d \; +$$

$$\ell_0^{\mu-1} \int_0^{\infty} \varepsilon \lambda_d (\xi_d - \lambda_d))\, h(\xi_d)\, d\xi_d$$

$$\approx \ell_0^{\mu-1} \lambda_d^{\varepsilon} + \mathcal{O}(\varepsilon) \tag{11}$$

where the approximation is based on a small variance assumption on the truncation length distribution, such that $|\xi_d - \lambda_d| \ll \lambda_d$. Replacing the value of $\varepsilon = 2 - \mu$ and inserting the integral approximation of Eq. (11) into Eq. (10), we arrive at a mean leg length formula for truncated flight searches of

$$\langle \ell \rangle \approx \left(\frac{1-\mu}{2-\mu} \right) \ell_0 + \left(\frac{1}{2-\mu} \right) \left(\frac{\lambda_d}{\ell_0} \right)^{1-\mu} \lambda_d + \mathcal{O}(|\mu - 2|). \tag{12}$$

Recalling that the mean leg length approximation formula is only valid for $\mu \approx 2$, we perform another regular asymptotic expansion of the expression in Eq. (12) via the substitution $\varepsilon = \mu - 2$, leading to the following:

$$\varepsilon \langle \ell \rangle \approx \ell_0 \left(\frac{\lambda_d}{\ell_0} \right)^{\varepsilon} - \ell_0\, (1 - \varepsilon) + \varepsilon\, \mathcal{O}(\varepsilon) \tag{13}$$

A first-order asymptotic expansion for $\varepsilon \langle \ell \rangle$ yields the following expression:

$$\varepsilon \langle \ell \rangle \approx \varepsilon \ell_0 \left[1 + \ln \left(\frac{\lambda_d}{\ell_0} \right) \right] + \mathcal{O}(\varepsilon^2) \tag{14}$$

which yields an expression for the achieved mean leg length of

$$\langle \ell \rangle \approx \ell_0 \left[1 + \ln \left(\frac{\lambda_d}{\ell_0} \right) \right] + \mathcal{O}(|\mu - 2|) \tag{15}$$

The expression in Eq. (15) shows that the mean leg length $\langle \ell \rangle$ is nearly linear in the minimum leg length ℓ_0, and that it increases with the mean value of the truncation length due to geometry (which is given by λ_d).

To find an analytic form for the mean truncation length λ_d, we turn to the literature on mean chord lengths in convex domains. In particular, the mean truncation length λ_d can be found from the mean chord length in a convex region $Q \subset \mathbb{R}^2$ according to [6]:

$$\lambda_d = \frac{\pi |Q|}{|\partial Q|} \tag{16}$$

where $|Q|$ is the area of the domain Q and $|\partial Q|$ is the perimeter. Note that for a circular domain of radius R this gives $\lambda_d = \pi R/2$, and for a rectangular domain of length D_L and width D_W this gives $\lambda_d = \pi D_L D_W/(2D_L + 2D_W)$ which yields $\lambda_d = \pi D/4$ for a square with sides of length D. The particular definition of *mean chord length* that is used to arrive at this formula for λ_d is for a set of random lines that intersect the convex domain. As pointed out in [9], the value of mean chord length has slight variations depending on the specific context under which the chords are generated. However, computer simulations of a variety of random search truncations in rectangular domains show the formula to hold well for the search truncation length application.

A set of Monte Carlo computer simulations of random flights in a bounded domain with truncations were performed to validate the formula given in Eq. (15) along with the formula for truncation length given by Eq. (16). The results of the computer simulations are shown in Fig. 2. We note that simulations show good agreement with the analytic form except for very large values of ℓ_0 (when $\ell_0 > 0.75\lambda_d$). In those cases, the minimum leg length of the distribution, ℓ_0, is so close to the mean truncation level, λ_d, that the effect of the distribution on the resulting legs is dominated by the effect of the truncations, which leads to the analytic expression underpredicting the simulated results. Fortunately, this is not a practical region of the domain, as demonstrated in the next section.

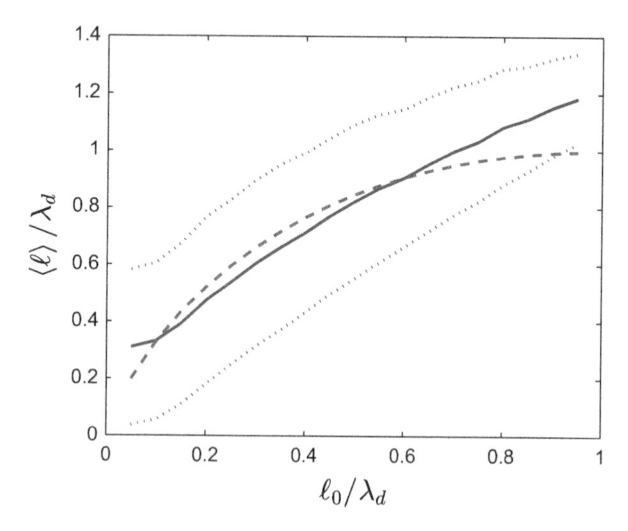

Fig. 2 Comparison of the analytic form for $\langle \ell \rangle$ given by Eq. (15) and Monte Carlo simulations of achieved lengths of truncated random search legs in a finite domain. The red dashed line is the analytic solution and the blue solid line is the mean of the simulation runs. The blue dotted lines show the mean plus/minus one standard deviation for the simulation results

3.2 Setting Search Parameter Values

In the previous section, the mean leg length for random flights in a truncated domain was determined and an analysis conducted showed that one can maximize detection rate by matching the expected mean free path between targets with the mean leg length. Next, we determine the parameter values for which that match holds true. In particular, for a mean free path between target detections given by λ_T, the mean free path and the mean leg length match (i.e., $\langle \ell \rangle = \lambda_T$) when the following holds true (given by substituting $\langle \ell \rangle = \lambda_T$ into Eq. (15)):

$$\frac{\ell_0}{\lambda_T} = \left[1 + \ln\left(\frac{\lambda_d}{\ell_0}\right)\right]^{-1} \tag{17}$$

which is an optimal search rule for setting the value of lower leg lengths ℓ_0 based on the domain size and expected target population.

The search leg parameter rule of Eq. (17) is difficult to implement in practical problems, since it involves solving a transcendental equation to find the value of ℓ_0. While a numerical solution may be practical for fixed individual situations, an approximate analytic form is desirable. Recognizing that values of $\lambda_T < \lambda_d$ are the only values of practical interest, we let $\alpha = \lambda_T/\lambda_d$ and find an approximate form for ℓ_0 under the range $0 \leq \alpha \leq 1$. From Eq. (17), we let $\xi = \ell_0/\lambda_d$ and arrive at

$$\alpha = \xi - \xi \ln(\xi). \tag{18}$$

To develop a rule for setting ℓ_0 from Eq. (18), we must invert the equation into a form $\xi = f(\alpha)$ (which becomes $\ell_0 = \lambda_d f(\lambda_T/\lambda_d)$ after substitution).

To form an analytically invertible approximation for Eq. (18), we develop a functional form that is polynomial in ξ and meets four primary features of Eq. (18), namely:

(1) $\alpha = 0$ when $\xi = 0$,
(2) $\alpha = 1$ when $\xi = 1$,
(3) $d\alpha/d\xi > 0$ for $\xi \in [0, 1]$,
(4) $d^2\alpha/d\xi^2 < 0$ for $\xi \in [0, 1]$.

A polynomial form that meets these features is given by:

$$\alpha \approx 1 - (1 - \xi)^{\beta} \tag{19}$$

for $\beta > 1$. Fitting integer values of β to the curve of $\alpha = \alpha(\xi)$, we find that the best integer match occurs for $\beta = 3$. A plot of the approximation of Eq. (18) using Eq. (19) with a value of $\beta = 3$ is shown in Fig. 3.

This approximation in Eq. (19) (with $\beta = 3$) is analytically invertible and leads to an approximation for ℓ_0 in terms of λ_d and α of the form

$$\ell_0 \approx \lambda_d \left[1 - (1 - \alpha)^{1/3}\right] \tag{20}$$

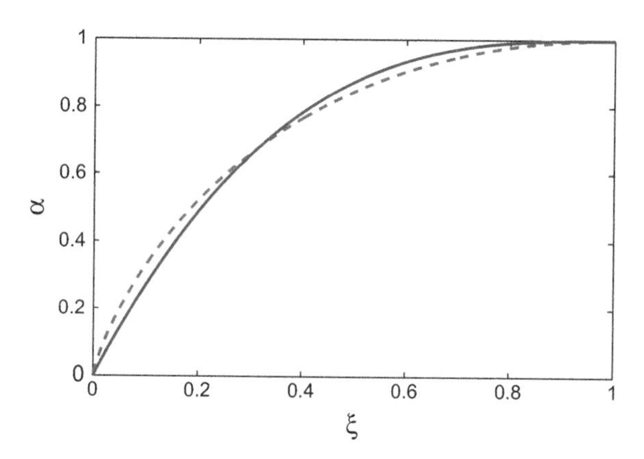

Fig. 3 Comparison of the exact and approximate form of α as a function of ξ. The red dashed line is the exact form of Eq. (18) and the blue solid line is the approximation from Eq. (19) with a value of $\beta = 3$

which in the physical parameters becomes

$$\ell_0 \approx \lambda_d \left[1 - \left(1 - \frac{\lambda_T}{\lambda_d} \right)^{1/3} \right] \tag{21}$$

We note that in the limit of very large domains, in particular as $\lambda_d \gg \lambda_T$, this approximation becomes

$$\ell_0 \approx \frac{\lambda_T}{3} \tag{22}$$

which shows the lower bound of the leg length distribution scales linearly with the size of the expected mean free path between targets as the size of the domain gets large, as expected. Equation (21) is the fundamental expression for the practical sizing of leg length parameters when performing random flights in bounded domains.

3.3 *Effect of Multiple Searchers*

We note that the addition of multiple random searchers now has the effect of merely decreasing the mean free path between target detections, as the search problem is known to have a duality between searcher and target frames of reference [21]. Thus, the extension to N searchers is accomplished by recognizing that $\lambda_T \mapsto \lambda_T/N$. In particular, for uniformly random objects distributed with density ρ in a two-dimension domain, the N-searcher mean free path becomes

$$\lambda_T = \frac{1}{2Nr_d\rho} \tag{23}$$

which is the equivalent to increasing the individual searcher detection range by a factor of N. This simple scaling only holds true for random non-collaborative searchers seeking randomly located objects, although many practical situations can be modeled in that way.

To see the effect of multiple searchers on the resulting search legs, we first simulated a search for a single ($N = 1$) searcher with a search length of $\sum_i L_i = L_{max}$ using the optimized search parameters. In this simulation, we applied a search of length $L_{max} = 10D$ over a square domain of size $D \times D$ with search range $r_d = D/20$. For this example, the resulting search effort of Eq. (2) for the total search is given by:

$$SE = \sum_i SE_i = 2r_d L_{max} = D^2 \tag{24}$$

This is the expected search coverage that would be achieved by such a searcher performing nonoverlapping search of the same length in the same domain. We note that $SE = |Q|$ in this case, showing that this searcher is expending the same amount of search effort that it would take to provide uniform 100% coverage of the domain if a path with no overlap could be achieved. The resulting path of the optimal random search of this searcher is given on the left of Fig. 4. On the right of Fig. 4, we show a search of $N = 4$ multiple searchers of the same capability (i.e., $r_d = D/20$) in the same domain, but with a search length of $L_{max} = 2.5D$. This was done so that the total search effort of all four combined searchers matches that of the individual searcher on the left. From these plots, it is clear that the multiple searchers have shorter "long legs" but still provide roughly an equivalent amount of search coverage as the individual searcher. Thus, a properly selected number of multiple independent (non-coordinated) searchers can be used to take the place of a single searcher if a reduction in individual total search length L_{max} is desired.

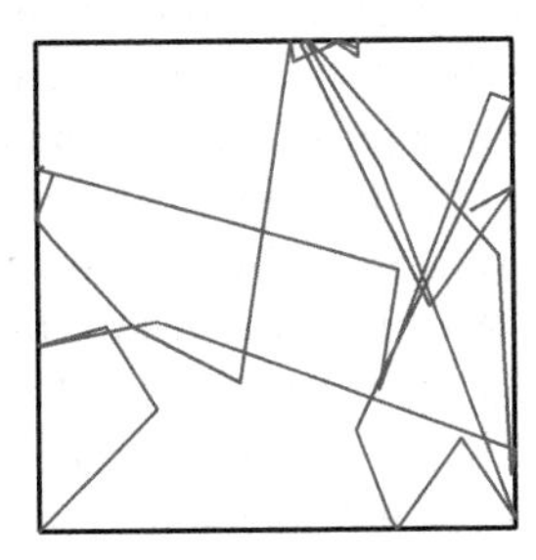

Fig. 4 Example of a search flight with one searcher for a fixed search distance D_0 (left) and an example of four searchers in the same domain with fixed search distances of $D_0/4$ with search parameters accounting for the $N = 4$ searchers

4 Numerical Examples

Summarizing the results of the previous sections, we have developed optimal search rules based upon a specific class of superdiffusive random flights whose search leg lengths are given by the probability distribution $p(\ell)$ as shown in Eq. (1). For a set of N searchers with detection range r_d following these random flight rules inside a bounded domain $Q \subset \mathbb{R}^2$, where the targets are distributed with density ρ, the optimal parameter for defining search leg lengths was shown to be given by:

$$\ell_0 \approx \frac{\pi |Q|}{|\partial Q|} \left[1 - \left(1 - \frac{|\partial Q|}{2\pi |Q| N r_d \rho} \right)^{1/3} \right] \tag{25}$$

which is found by substitution of Eqs. (16) and (23) into Eq. (21).

We first simulate the performance of a single ($N = 1$) searcher in a square domain Q of size $D \times D$. The objects of search are randomly distributed in the domain at a density of $\rho = 25/D^2$, and the searcher detects an object whenever it comes within a distance $r_d = D/20$ of the object. We simulate two types of search, the first is a mow-the-lawn style search of total length $L_{max} = 2.5D$ and the second is an optimized Lévy-style flight search of the same total search length. We note that this search length yields an expected coverage of $0.25D^2$, corresponding to 25% of the domain. Since the objects are randomly distributed, we expect that, on average, the mow-the-lawn search plan should yield detection of 25% of the objects. We ran 500 simulations of this scenario, where for each run a new set of search object locations were distributed and a new random search plan was generated (note that the mow-the-lawn plan, being a deterministic strategy, was the same for each run). All searches were initiated from the lower-left corner of the domain, and a sample showing four of the generated random search plans is shown in Fig. 5.

The resulting search effectiveness of the search approaches was measured by the number of objects found in each search. An object is considered to be found if the search path comes within a distance r_d of the object at any time during the search. To normalize, we record the percentage of objects that are found (number of found objects divided by the total number of objects in the domain) and these percentages were compiled into histograms for each of the search strategies. The resulting search performance percentages are given in Fig. 6. From these histograms, it is clear that the mow-the-lawn generally finds slightly less than 25% of the objects. This is because the need to "turn around" between legs of the mow-the-lawn strategy has some reduction in efficiency. Furthermore, due to the randomness of the object deployments, there are some cases when more or less than 25% are found. On the right side of the figure, it is clear that the optimal Lévy flight search achieves nearly the same performance as the mow-the-lawn search. This shows that the optimality achieved, while only guaranteed to be optimal among random flight strategies, still achieves performance nearing that of the optimal planned deterministic search.

Given that the optimized Lévy flight search has shown comparable performance to the structured mow-the-lawn search for a problem that is fairly sparse, we develop

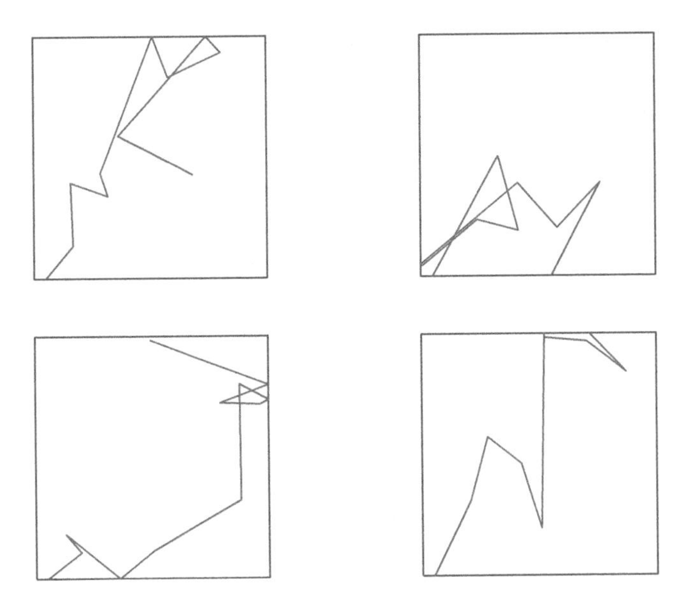

Fig. 5 Four examples of Lévy flight search paths generated for the numerical simulations

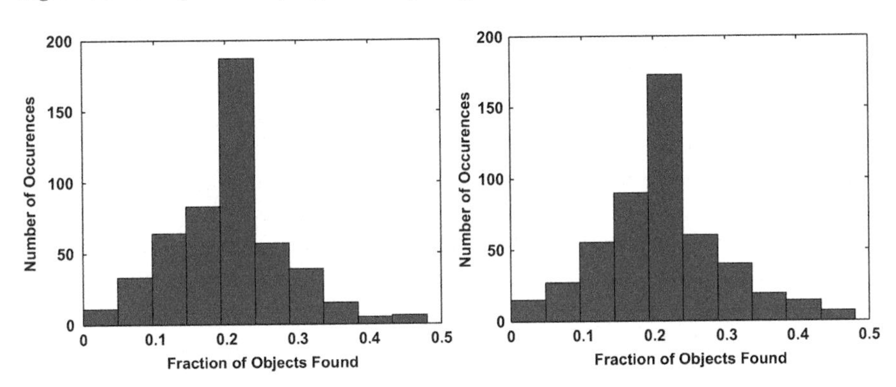

Fig. 6 Histograms of number of simulation runs (out of 500) where the search successfully found varying fractions of the total number of objects. The left histogram is for a structured mowing-the-lawn search and the right histogram is for an optimized Lévy flight search

next a numerical example to assess the sparsity requirement. We consider again a comparison of a mow-the-lawn search to an optimized Lévy flight search on the same domain. Once again, we consider the scenario to consist of a domain Q that is square of size $D \times D$ with objects placed randomly with density $\rho = 25/D^2$ and a detection range of $r_d = D/20$. To vary the sparsity of the scenario, we consider different search lengths ranging from $L = 0.1 \cdot L_{max}$ to $L = L_{max}$, where L_{max} corresponds to $SE = |Q|$ (i.e., the search length of a complete coverage search). For each search length, we run 100 scenarios with both the mow-the-lawn search plan and the optimized Lévy flight search. Rather than compare raw numbers of objects

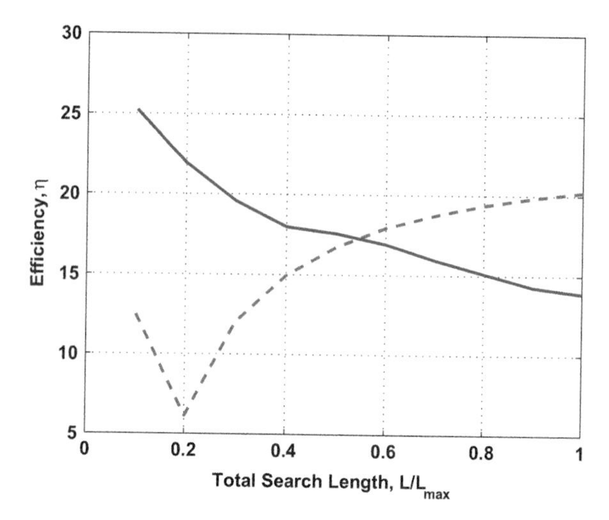

Fig. 7 The search efficiency η (as per Eq. (3)) for search by mow-the-lawn search planning (dashed red curve) and for optimized bounded Lévy flight search (solid blue curve) over searches of increasing total lengths. Each point is an average of 100 simulations

found, in this case we examine the resulting search efficiency, η, as given in Eq. (3). Recall that search efficiency is the ratio of the number of object detections to the total search effort applied by a searcher. Thus, it is a measure of the incremental benefit gained by adding an increment of search length.

The results of the search efficiency simulations as a function of search length are given in Fig. 7. These curves represent the average efficiency for each search strategy taken over the 100 scenario runs. From this figure, there are a couple of primary observations. First, for sparse search situations (small search length) the random flight search is clearly more efficient than the mow-the-lawn strategy. We note a small uptick in mow-the-lawn search for very short searches, this is due to added efficiency when there is a single leg of mow-the-lawn and no turnarounds are required. Second, as the search length increases, the random flight search becomes less efficient while the mow-the-lawn search becomes more efficient, until eventually the mow-the-lawn search efficiency surpasses that of the random search. Thus, the mow-the-lawn search is desirable when there is sufficient search effort applied; yet, if there is not much effort applied (or there is a desire for rapid finding of objects, as in search-and-rescue), the optimized Lévy flight search is preferable. This matches with the intuition from animal foraging scenarios that occurred in unbounded domains, and thus by optimizing the random flight for the bounded domain, we maintain the qualitative benefits seen in unbounded search.

5 Conclusion

We have developed a new method for computing the parameters defining an optimally efficient random flight search for searchers operating in a bounded domain. By building on the work of modeling the behavior of animal foraging, we have created a method to set parameters for searches based on the specifics of the scenario. The search results have benefits for sparse search scenarios, and have been shown to achieve performance efficiency that surpasses optimal coverage search (i.e., mow-the-lawn) in sparse scenarios. This type of random search plan has benefits in that it does not require the navigation precision of a planned path search, and it also is readily adapted to scenario changes, as each leg is determined independently of past legs. We also have shown how to apply the method to multiple non-collaborative searchers operating in the same domain. The use of this approach for search planning is particularly beneficial for searchers operating with limited navigational accuracy, an example of which is unmanned vehicles operating in the marine environment. In addition to their potential use as planning algorithms on search platforms, these algorithms provide an optimal performance for a limiting type of search that can be used to benchmark system performance.

References

1. M. Andersson. On optimal predator search. Theor. Popul. Biol. **19**(1), 58–86 (1981)
2. B.W. Andrews, K.M. Passino, T.A. Waite, Foraging theory for autonomous vehicle decision-making system design. J. Intell. Robot. Syst. **49**, 39–65 (2007)
3. B.W. Andrews, K.M. Passino, T.A. Waite, Social foraging theory for robust multiagent system design. IEEE Trans. Autom. Sci. Eng. **4**(1), 79–86 (2007)
4. F. Bartumeus, J. Catalan, U.L. Fulco, M.L. Lyra, G.M. Viswanathan, Optimizing the encounter rate in biological interactions: Lévy versus Brownian strategies. Phys. Rev. Lett. **88**(9), Article number 097901 (2002)
5. D. Boyer, O. Miramontes, H. Larralde, Lévy-like behaviour in deterministic models of intelligent agents exploring heterogeneous environments. J. Phys. A: Math. Theor. **42**, Article number 434015 (2009)
6. N. Chernov, Entropy, Lyapunov exponents, and mean free path for billiards. J. Stat. Phys. **88**(1–2), 1–29 (1997)
7. N.E. Humphries, N. Quieroz, J.R.M. Dyer, N.G. Pade, M.K. Musyl, K.M. Schaefer, D.W. Fuller, J.M. Brunnschweiler, T.K. Doyle, J.D.R. Houghton, G.C. Hays, C.S. Jones, L.R. Noble, V.J. Wearmouth, E.J. Southall, D.W. Sims, Environmental context explains Lévy and Brownian movement patterns of marine predators. Nature **465**, 1066–1069 (2010)
8. B.O. Koopman, *Search and Screening: General Principles with Historical Applications* (Pergamon Press, New York, 1980)
9. P.W. Kuchel, R.J. Vaughan, Average lengths of chords in a square. Math. Mag. **54**(5), 261–269 (1981)
10. M. Mangel, C. Clark, Search theory in natural resource modeling. Nat. Resour. Model. **1**, 1–54 (1986)
11. V.V. Palyulin, A.V. Chechkin, R. Metzler, Lévy flights do not always optimize random blind search for sparse targets. Proc. Natl. Acad. Sci. **111**(8), 2931–2936 (2014)
12. E.P. Raposo, S.V. Buldyrev, M.G.E. da Luz, G.M. Viswanathan, H.E. Stanley, Lévy flights and random searches. J. Phys. A: Math. Theor. **42**, Article number 434003 (2009)
13. G.D. Ruxton, Increasing search rate over time may cause a slower than expected increase in prey encounter rate with increasing prey density. Biol. Lett. **1**, 133–135 (2005)
14. M.C. Santos, E.P. Raposo, G.M. Viswanathan, M.G.E. da Luz, Optimal random searches of revisitable targets: crossover from superdiffusive to ballistic random walks. Europhys. Lett. **67**(5), 734–740 (2004)

15. J.M. Smith, G.R. Price, The logic of animal conflict. Nature **246**, 15–18 (1973)
16. D.W. Stephens, J.R. Krebs, *Foraging Theory* (Princeton University Press, Princeton, 1986)
17. J.M.J. Travis, S.C.F. Palmer, Spatial processes can determine the relationship between prey encounter rate and prey density. Biol. Lett. **1**, 136–138 (2005)
18. G.M. Viswanathan, V. Afanasyev, S.V. Buldyrev, E.J. Murphy, P.A. Prince, H.E. Stanley, Lévy flight search patterns of wandering albatrosses. Nature **381**, 413–415 (1996)
19. G.M. Viswanathan, S.V. Buldyrev, S. Havlin, M.G.E. da Luz, E.P. Raposo, H.E. Stanley, Optimizing the success of random searches. Nature **401**, 911–914 (1999)
20. G.M. Viswanathan, V. Afanasyev, S.V. Buldyrev, S. Havlin, M.G.E. da Luz, E.P. Raposo, H.E. Stanley, Lévy flights in random searches. Phys. A **282**, 1–12 (2000)
21. T.A. Wettergren, C.M. Traweek, The search benefits of autonomous mobility in distributed sensor networks. Int. J. Distrib. Sens. Netw. **8**(2) (2012). Article ID 797040
22. K. Zhao, R. Jurdak, J. Liu, D. Westcott, B. Kusy, H. Parry, P. Sommer, A. McKeown, Optimal Lévy-flight foraging in a finite landscape. J. R. Soc. Interface **12**, Article number 20141158 (2015)

Index

A. A. Ruffa, B. Toni (eds.), *Advanced Research in Naval Engineering*, STEAM-H: Science, Technology, Engineering, Agriculture, Mathematics & Health, https://doi.org/10.1007/978-3-319-95117-1

W

Printed in the United States
By Bookmasters